Petrology of the metamorphic rocks

Second Edition

Roger Mason

London
UNWIN HYMAN
Boston   Sydney   Wellington

Published by the Academic Division of
Unwin Hyman Ltd
15/17 Broadwick Street, London W1V 1FP, UK

Unwin Hyman Inc.
955 Massachusetts Avenue, Cambridge, MA 02139, USA

Allen & Unwin (Australia) Ltd
8 Napier Street, North Sydney, NSW 2060, Australia

Allen & Unwin (New Zealand) Ltd
in association with the Port Nicholson Press Ltd
Compusales Building, 75 Ghuznee Street, Wellington 1, New Zealand

First published in 1990

British Library Cataloguing in Publication Data
Mason, Roger 1941–
    Petrology of the metamorphic rocks. – 2nd ed.
1. Metamorphic rocks. Petrology
I. Title
552.4

ISBN 0–04–552027–5
ISBN 0–04–552028–3   pbk

Library of Congress Cataloging-in-Publication Data
Mason, Roger, 1941–
    Petrology of the metamorphic rocks/Roger Mason. — 2nd ed.
        p.   cm.
Includes bibliographical references.
ISBN 0–04–552027–5. — ISBN 0–04–552028–3 (alk. paper)
1. Rocks, Metamorphic.   I. Title.
QE475.A2M394   1990
552'.4—dc20   90–35106
                                                            CIP

Typeset in 10 on 12 point Times by Computape (Pickering) Ltd,
North Yorkshire
and printed in Great Britain by Cambridge University Press

Preface

There has been a great advance in the understanding of processes of metamorphism and of metamorphic rocks since the last edition of this book appeared. Methods for determining temperatures and pressures have become almost routine, and there is a wide appreciation that there is not a single temperature and pressure of metamorphism, but that rocks may preserve, in their minerals, chemistry and textures, traces of their history of burial, heating, deformation and permeation by fluids. However, this exciting new knowledge is still often difficult for non-specialists to understand, and this book, like the first edition, aims at enlightenment. I have concentrated on the interpretation of the plate tectonic settings of metamorphism, rather than following a geochemical approach. Although there is an impressive degree of agreement between the two, I believe that attempting to discover the tectonic conditions accompanying rock recrystallization will more readily arouse the interest of the beginner.

I have used a series of case histories, as in the first edition, drawing on my own direct experience as far as possible. This means that some subjects are treated in more detail than others, and many important topics are barely mentioned at all. It also means that general concepts appear in a rather haphazard order in the text. To help my readers, I have provided a glossary of definitions of terms used in the book, which are indicated in **bold type** in the text. I discuss thermotectonic models, based on very simple 2-dimensional analyses, in describing the metamorphic history of many of my examples. Although many problems remain to be solved in this approach (and I do not claim that my models are sufficiently rigorous to be contributions to research discussion), its success in accounting for the shape of many P–T–t paths has convinced me that it is a useful framework to use in learning about metamorphism. I hope I have provided my readers with an interesting and sound basis for further exploration of this fascinating branch of the Earth sciences.

The present form of the book arose from a course of lectures and demonstrations given at Wuhan College of Geology (since renamed the China University of Geosciences) in the People's Republic of China in 1986. I thank President Zhao Pengda and Professor You Zhendong for inviting me to give the course, Mrs Han Yuqing who worked very hard as my interpreter, and the students of the class who suffered the culture shock of being introduced to western-style discussion in lectures! I have been encouraged to emphasize the tectonic theme in a thorough and constructive review of my manuscript by Dr M. J. Le Bas. In addition to those I thanked for advice in my first edition, I would like to thank Dr L. Ashwal, Dr A. P. Boyle, Professor K. Burke, Dr K. W. Burton, Mr A. J. Griffiths, Dr W. L. Kirk, Dr J. D. A. Piper, Dr G. D. Price, Ms K. V. Wright, Dr J. W. Valley, and Dr C. Xenophontos. I thank Dr P. W. Edmondson for compiling the index. I also acknowledge with thanks receipt of a travel grant from the Central Research Fund of the University of London to visit classic metamorphic localities in the USA, and also a grant from the Royal Society of London to the Sulitjelma '83 Ophiolite Expedition.

Contents

List of tables

For Marion

1 Introduction

Definition

There are three categories of rocks – igneous, sedimentary and metamorphic. Igneous rocks have been formed by solidification of hot silicate melts known as magmas, which may be erupted at the Earth's surface as lavas from volcanoes. Sedimentary rocks form by a variety of processes at low temperatures near the surface, including the bottom of the seas and oceans. Metamorphic rocks, the third category, were originally either igneous or sedimentary, but their character has been changed by processes operating below the surface of the Earth. The constituent minerals of the rocks may be changed, or the shapes, sizes and mutual spatial relationships of the crystals may change. There are a variety of processes involved, and they are collectively described by the term **metamorphism**.

The most important factor controlling metamorphic processes has been found to be temperature. Because heat energy is produced within the Earth by the decay of radioactive **isotopes**, temperature usually increases with depth, and so the more deeply rocks have been buried, the more they are likely to have been affected by metamorphism. The significance of temperature in metamorphism allows bounds to be set upon the range of conditions involved; if temperatures are sufficiently low to be near those of the surface, the processes are regarded as sedimentary, if they are sufficiently high for silicate melts to form, they are said to be igneous. More precise definitions of these limits will be left until later in this book.

From the limits to metamorphism described above, it follows that metamorphism takes place when the rock-forming minerals concerned are solid. Fluids may be involved, but they will contain a percentage of water or be rich in other substances (such as carbon dioxide) which are liquids or gases at the surface of the Earth. The processes of metamorphism are therefore *solid state* processes, with or without the involvement of volatile fluids.

As mentioned earlier, the minerals in a rock may be changed by metamorphism, with original igneous or sedimentary minerals disappearing and new metamorphic minerals appearing. Or the alteration may be less extreme, with changes in the relative proportions of different minerals, or in the chemical compositions of minerals. Changes of this type are described as changes in the **mineral assemblage** of a metamorphic rock. There may also be changes in the sizes, shapes and mutual relationships of the minerals of the rock. These are described as changes in the **fabric** or **texture** of the rock.

The definition of metamorphic rocks may therefore be formulated as follows. *Metamorphic rocks are those whose mineral assemblages and/or textures have been altered by solid state processes operating within the Earth, above the temperatures usually found at the surface, but below the temperatures at which silicate melts form.*

The processes of metamorphism

As stated above, temperature can be shown to be the most important influence on the change of rocks during metamorphism. A metamorphosed sediment, for example, will have been deposited at a low temperature (about the same as the present temperature at the Earth's surface), undergone a period of heating, then a period of cooling, until it was available for collection and study in its present form (Fig. 1.1).

Pressure also changes during metamorphism. Pressure would increase due to the increasing load on the sediment as younger sediments are deposited above it. Then the pile of sediments might be uplifted and eroded, so that pressure fell until the specimen was at the surface, under negligible pressure as far as metamorphism is concerned. Figure 1.2 shows a cycle of increasing pressure, then decreasing pressure caused by such burial and uplift.

However, it is likely that both pressure and temperature will change, because as rocks are buried beneath the Earth's surface they become heated, due to the blanketing effect of the pile of poorly conducting sediments above them. The heating and cooling history of a rock might be illustrated by two graphs, both on the same time axis, one showing temperature, the other

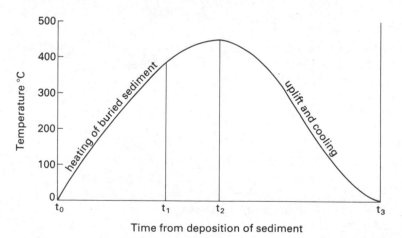

Figure 1.1 Temperature–time plot for a sedimentary rock which has undergone a single cycle of metamorphism. t_0–time of deposition of sediment, t_1–time of maximum pressure, t_2–time of maximum temperature, t_3–end of metamorphism and uplift (rock could be collected at the surface).

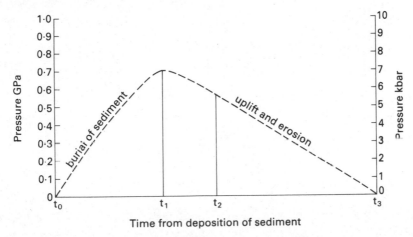

Figure 1.2 Pressure–time plot for the sedimentary rock shown in Figure 1.1. t_0, t_1, t_2, t_3 as in Figure 1.1.

showing pressure (Fig. 1.3). The accumulating pile of sediment takes time to heat up, and so there will be a time lag between increase of pressure and increase of temperature, which is shown in Figure 1.3 by the maxima in the pressure and temperature curve being reached at different times (t_1 and t_2). An alternative way of expressing this relationship is to plot pressure against temperature. Because temperature and pressure do not increase or decrease in step with one another, the process of metamorphism will be expressed as a loop on the pressure–temperature (PT) graph, and arrows can be drawn on the loop to indicate how conditions changed during time (t) (Fig. 1.4). Such a plot is called a pressure–temperature–time (P–T–t) loop. The example discussed here is a very simple one, and more complicated **P–T–t paths** are obviously likely to occur.

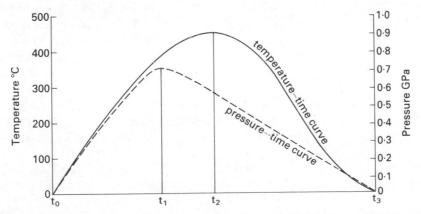

Figure 1.3 Temperature and pressure graphs from Figure 1.1 and Figure 1.2, combined on the same diagram. t_0, t_1, t_2, t_3 as in Figure 1.1.

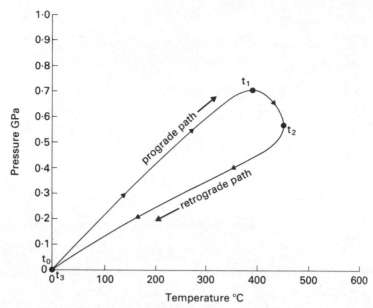

Figure 1.4 Pressure–temperature–time (P–T–t) loop, showing the temperature and pressure changes of Figure 1.1 and Figure 1.2. t_0, t_1, t_2, t_3 as in Figure 1.1.

Figure 1.4 is plotted in the orientation familiar to metamorphic geologists, with temperature as the horizontal abscissa, and pressure as the vertical ordinate. However, because pressure increases with depth, petroleum geologists and geophysicists use an alternative plot, with depth as the horizontal axis, and temperature as the vertical one (Fig. 1.5). If the sedimentary rocks filling a basin are weak, it is possible to convert depth into pressure, using a series of assumptions and equations which will be presented in a later section of this book (Fig. 1.6). A variety of units are used to record pressures in rocks; published papers often use bars and kilobars, but the Standard International (S.I.) unit of pressure, the pascal (Pa), which is equal to 1 newton per square metre ($N\ m^{-2}$) is increasingly appearing as well. Table 1.1 gives the equivalent values of units used for pressure, in pascals and in gigapascals ($1\ GPa = 10^9\ Pa$), which are conveniently large units for describing pressure in the Earth's crust and upper mantle.

It is possible to obtain P–T–t curves in two ways; by thermal modelling to arrive at a calculated curve, and by study of the metamorphic minerals present in a metamorphic rock to estimate a number of temperatures and pressures, for different times in the metamorphic history. The latter technique is known as **geothermobarometry**. The maximum temperatures reached may be found at different places in a region of metamorphic rocks and the information used to construct a **field metamorphic gradient**. Provided that the rocks remained at the same depths for a long time, this gradient will correspond to the increase in temperature with depth in part of the Earth's

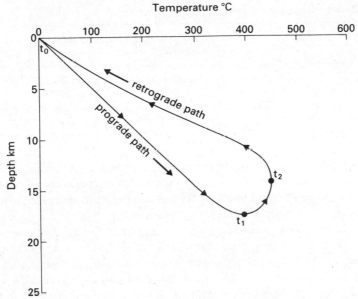

Figure 1.5 Depth–temperature–time loop, showing the same curves as in Figure 1.1 and Figure 1.2. t_0, t_1, t_2, t_3 as in Figure 1.1.

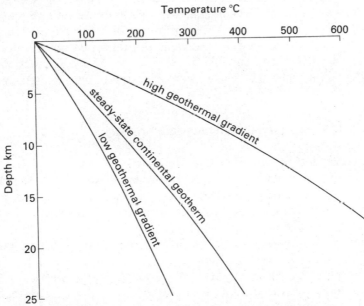

Figure 1.6 Depth–temperature curve for stable continental lithosphere, and for areas of high and low geothermal gradients.

Table 1.1 Pressure units used in petrology

Unit name	Symbol	Equivalent value	
		pascals	gigapascals
atmosphere	atm	1.013×10^5 Pa	1×10^{-4} GPa (approx.)
bar	bar	1×10^5 Pa	1×10^{-4} GPa
kilobar	kbar	1×10^8 Pa	0.1 GPa
pound per square inch	p.s.i.	6.895×10^3 Pa	6.895×10^{-6} GPa

crust at a particular time. Temperature/depth curves are known as geothermal curves or **geotherms**, and the rate of increase of temperature with depth is known as the **geothermal gradient**. If the temperature increases rapidly with depth, the geothermal gradient is described as 'high', if it increases slowly it is described as 'low' (Fig. 1.5). Notice that this creates a presentation problem on the temperature/pressure (Fig. 1.6), and temperature/depth graphs. The gradient of a geothermal curve on such a graph is *low* for a high geothermal gradient, and *high* for a low geothermal gradient. This is a result of the way the axes are arranged, (and causes beginners a lot of confusion).

Modern research is concerned with obtaining P–T–t curves by geothermobarometry, comparing P–T–t curves across regions of metamorphic rocks to obtain metamorphic pressure–temperature gradients, and applying the results to plate tectonic models. The method gives an insight into tectonic processes which is not obtainable in other ways. Many assumptions are involved, and this book will review them in a series of particular examples.

The three field categories of metamorphic rocks

The fundamental definition of metamorphic rocks, given above, is a genetic one, i.e. it is based upon the mode of origin of the rocks. The broad subdivision of metamorphic rocks is also genetic, using the field relationships of the rocks concerned to divide them into three classes, contact, dynamic and regional metamorphic rocks.

Contact metamorphic rocks occur over restricted areas of the Earth's surface, near the *contacts* with igneous intrusions. In many cases, the degree of metamorphism can be seen to be greatest when in contact with the igneous rocks, and to die away as distance from the contact increases. This relationship shows that the main agency causing metamorphism is the heat escaping from the hot magma and continuing to escape from the hot solid rocks into the surrounding **country rocks** after the igneous intrusion has solidified. For this reason, contact metamorphic rocks are often called thermal metamorphic rocks.

Dynamic metamorphic rocks are also found over restricted areas of the Earth's surface, close to major faults or thrusts. This relationship, and many features of the textures of the rocks, indicate that the rocks were formed by the concentrated deformation occurring near the fractures. One important process concerned is **cataclasis**, the mechanical breaking of the mineral grains in the rock, and dynamic metamorphic rocks are therefore sometimes called **cataclastic metamorphic rocks**. This name is misleading because cataclasis is only one of the processes concerned, and in many cases not the most significant in bringing about the change in mineral assemblage and texture.

Regional metamorphic rocks occur over much larger areas than contact or dynamic metamorphic rocks, constituting most of the continental crust, and much of the oceanic crust and the upper mantle. Their textures and field relationships indicate that the processes of metamorphism were closely linked with the formation and deformation of crust and mantle rocks, generally involving both heating and deformation. For this reason regional metamorphic rocks are sometimes known as **dynamothermal metamorphic rocks**.

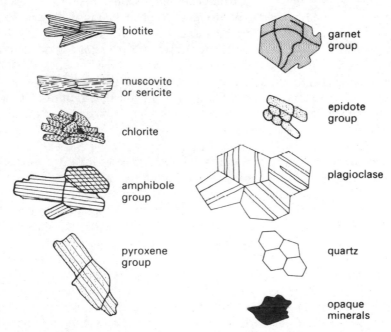

Figure 1.7 Common ornaments used for different types of minerals in the thin section drawings.

Studying thin sections of metamorphic rocks

This textbook is designed for readers who have the use of a petrological microscope and thin sections of metamorphic rocks. A series of case histories are discussed, including descriptions of individual thin sections from metamorphic rock suites, emphasizing the features which can be recognized in the thin sections. Some hints on microscopic technique are included where appropriate, but this section makes some general comments on the use of the petrological microscope for metamorphic rocks. For systematic descriptions of the optical properties of the minerals which are found in metamorphic rocks, the reader is referred to the textbooks of Gribble & Hall (1985) and Deer *et al.* (1966).

In general, the features discussed are those displayed by transparent minerals, which can be identified in thin section. Opaque minerals are usually described as 'opaques'. The textural features are illustrated by line drawings, using a set of conventional ornaments (Fig. 1.7) for common mineral groups, but some photomicrographs are also included. The aim of this is to show how the conventional drawings relate to the image actually seen when looking down a petrological microscope. Ideally, this topic should be illustrated by colour photomicrographs taken in plane polarized light and between crossed polars, but since it is textural features rather than optical properties which are usually the subject of discussion, colour photographs have been omitted in order to keep the price of the book down. Study the photomicrographs in McKenzie & Guildford's 'Atlas of the rock-forming minerals in thin section' (1980), especially the illustrations of minerals such as kyanite, diopside, wollastonite and hornblende, for excellent colour illustrations of the appearance of typical metamorphic minerals in metamorphic rocks.

To improve your ability to identify minerals and record textures, make your own drawings of thin sections, showing all the minerals in the rock's mineral assemblage. It is useful to include quite a large number of individual crystals, showing their actual outlines as accurately as possible, so that the fabric of the metamorphic rock is displayed. Try to avoid the mistake of filling in the finer-grained **groundmass** between larger **porphyroblasts** with shading or an ornament which bears little relation to the actual shapes and arrangement of smaller crystals. It is usually best to draw metamorphic rocks as they appear in plane polarised light, even though you may have distinguished some minerals between crossed polars (e.g. quartz from plagioclase feldspar), because features such as birefringence, or twinning, which are more clearly seen between crossed polars, seldom reflect the metamorphic processes involved in the evolution of the rock, although they may be crucial for mineral identification.

Most students begin the study of metamorphic rocks when they are already quite experienced with igneous and sedimentary rocks. This means that you should be able to recognise any primary features which survive from the parent igneous or sedimentary rocks. In studying metamorphic rocks, it is important to bring from igneous petrology the practice of always identifying all minerals in the rock, and from sedimentary petrology careful attention to grain size, grain shape and other textural features. You should be prepared to identify minerals even when they occur as small crystals, and to study the relationships between crystals and the shapes of crystal boundaries under high magnifications.

Laboratory study of metamorphic rocks

There has been a remarkable increase in recent years in the range and capabilities of laboratory methods of all kinds, as applied to metamorphism and to metamorphic rocks. While the central technique discussed in this book is the traditional one of study of thin sections of rocks under the petrological microscope, other techniques will also be referred to. Brief outlines of important techniques will be given, but they are only intended as introductions, and apply only from a metamorphic point of view. If you would like to know more about laboratory methods, please refer to the more specific texts, which are mentioned in the appropriate places.

Radiometric dating of metamorphic rocks

We have seen that the decay of radioactive isotopes produces heat energy within the Earth, but the main use of the measurement of radioactive isotopes in geology is for age dating. Much of this has been done on metamorphic rocks, especially those of Precambrian age. An understanding of the processes involved in metamorphism has been found to be essential for correct interpretation of the dates obtained, but conversely radiometric studies whose primary aim is dating also yield valuable insights into metamorphism (Faure 1986).

The unstable isotopes of five elements are commonly used for radiometric age dating. They are potassium 40 (^{40}K), rubidium 87 (^{87}Rb), samarium 147 (^{147}Sm), thorium 232 (^{232}Th), uranium 235 (^{235}U) and uranium 238 (^{238}U). Potassium is the most abundant element in the list and although ^{40}K only makes up a small proportion of the potassium in the earth, it is the most abundant radioactive isotope. Because potassium is an essential element in micas and clay minerals, it is found in appreciable quantities in many igneous and sedimentary rocks. Similarly, rubidium is present in many rocks at trace element concentrations, usually of the order of ten parts per million (p.p.m.). Samarium is a rare earth element (REE), present in very small concentrations, but of considerable interest in metamorphism because it is concentrated into some minerals, including epidote and garnet, which are

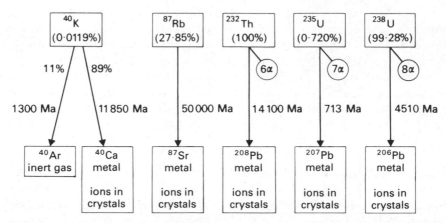

Figure 1.8 Radioactive decay of the principal unstable isotopes used to date metamorphic rocks. The upper boxes give the chemical symbols of the parent isotopes, and the percentage of the atoms of the naturally occurring element which are of each parent isotope. The lower boxes show the final stable daughter isotopes, and whether they are unreactive gas (which is not incorporated into the structure of rock-forming minerals) or metallic ions (which can be incorporated into the minerals).

widely formed by metamorphic processes. Uranium and thorium are similarly rare elements, but tend to be concentrated into accessory minerals such as zircon, apatite and orthite, which are found in many metamorphic rocks.

The natures of the **daughter isotopes**, which are the end products of radioactive decay of **parent isotopes**, are crucial for the interpretation of the relative concentrations of parent and daughter isotopes in metamorphic rocks, which are used to determine ages of metamorphism. Figure 1.8 shows the daughter isotopes produced by the decay of the parent isotopes shown above them on the diagram listed above, and also gives the **half lives** for each decay sequence. In the cases of uranium and thorium isotopes, only the first and last isotopes at either end of complex decay sequences are given.

Argon, the element formed by the decay of ^{40}K, is an inert gas which does not combine chemically with minerals. At low temperatures, ^{40}K ions are trapped in the crystal lattices of minerals, but the gas is rather easily driven out by heating. By contrast, strontium is a metallic element, which can be incorporated into the crystal structure of some minerals such as plagioclase feldspar, by the substitution of Sr^{+2} for Ca^{+2}. Therefore, on heating, ^{87}Sr tends to be retained in minerals growing during metamorphism, rather than to diffuse out of the rock in the same manner as ^{40}K. U and Th are present in zircon crystals, which are very resistant to solution in magmatic or metamorphic fluids, so that zircons may retain ages of magmatic crystallization of igneous rocks surviving from a period before metamorphism.

Potassium–argon age determinations measure the time since the rock or mineral analysed began to retain ^{40}Ar produced by the decay of ^{40}K, rather than the gas escaping by diffusion. Since diffusion is a thermally activated process, the retention of Ar decreases rapidly once a critical temperature, known as the **blocking temperature**, is exceeded (Dodson 1973). Because the blocking temperatures for different minerals vary, potassium–argon dating applied to different minerals separated from the same rock may give different results. If cooling from the maximum temperature on the P–T–t path was slow, minerals with higher blocking temperatures will yield higher ages than minerals with low blocking temperatures (Fig. 1.9). Complications arise if a rock has been reheated after initial cooling from a high temperature. If the later period of heating comes into the blocking temperature range there may be partial loss of argon, giving unreal 'ages' for the rock or minerals, intermediate between the age of the first cooling and the age of reheating.

There are various methods for overcoming this problem, and one will be described here. The rubidium–strontium 'clock' is less readily disturbed by

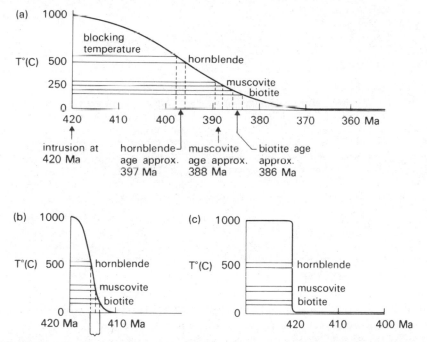

Figure 1.9 The relationship betwen potassium–argon ages in separated minerals from the same rock and the rate of cooling of the rock. (a) This rock cooled slowly. Hornblende, muscovite and biotite give different ages of 397, 388 and 386 Ma respectively, because of their different blocking temperatures. This curve is characteristic of the cooling history of regional metamorphic rocks. (b) This rock cooled during an interval of 10 Ma. Although the hornblende, muscovite and biotite should give different ages, the differences are smaller than the experimental error in potassium–argon dating. This curve is characteristic of rocks which underwent regional metamorphism at shallow levels in the lithosphere. (c) This rock cooled rapidly over an interval of a few thousand years. This curve is characteristic of rocks from high-level contact aureoles.

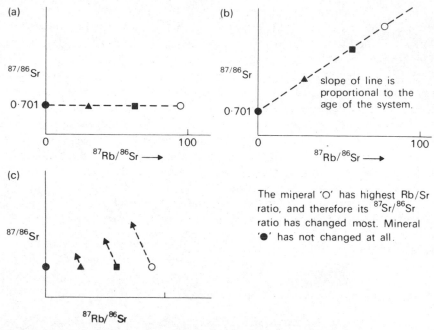

Figure 1.10 Evolution of isotopic ratios on isochron diagrams. $^{87}Sr/^{86}Sr$ ratios are plotted along the horizontal axis, $^{87}Rb/^{86}Sr$ ratios along the vertical axis. The closed circles show the ratios in a mineral containing no Rb, the triangle, square and open circle the ratios in minerals with increasing Rb contents. (a) Ratios at the time of crystallization of the rock. (b) Ratios after several million years. The mineral with the highest Rb/Sr ratio (open circle) shows the largest change in its $^{87}Sr/^{86}Sr$ ratio, while the mineral with no Rb shows no change in the ratio. The points fall on a straight line whose slope is proportional to the time since the minerals ceased to crystallize. (c) The dotted lines with the arrows show the change in the isotopic ratios of the individual minerals during this period of time.

reheating after primary cooling than the potassium–argon clock. Rb : Sr dates on *minerals* give the time which has elapsed since Rb and Sr ceased to be able to diffuse freely through the rock sample and were concentrated into the different minerals in different concentrations. A proportion of ^{87}Sr was already present when the rocks originally crystallized, so it is necessary to study the increase in the proportion of ^{87}Sr as well as the reduction in the proportion of ^{87}Rb in the minerals and in the whole rock. This is done by preparing **isochron diagrams**, which plot the ratio of ^{87}Sr to ^{86}Sr (a stable isotope) vertically, and the ratio of ^{87}Rb to ^{86}Sr horizontally (Fig. 1.10). Because some minerals have a higher ratio of Rb to Sr than others (some contain no Rb at all), the final ^{87}Sr : ^{86}Sr ratios will be higher in the Rb-rich minerals.

It is possible to plot isochron diagrams for individual rocks using several different minerals with different initial ratios of Rb to Sr. The isotope ratios should define a straight line on the diagram whose slope gives a measure of

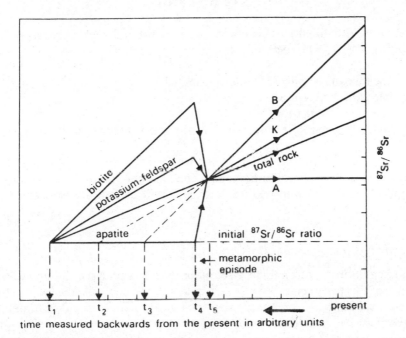

Figure 1.11 Isochron diagram for the minerals and whole rock in a granitic rock which underwent metamorphism after its magmatic crystallization. The metamorphism occurred from time t_4 to time t_5, and caused recrystallization of all the minerals in the rock, making their isotopic ratios of ^{87}Sr and ^{86}Sr the same, although they kept their different Rb/Sr ratios. It is also assumed that Sr did not diffuse into or out of the rock during metamorphism (i.e. there was no metasomatism). If the initial $^{87}Sr/^{86}Sr$ ratio can be calculated, it is possible to calculate the time of intrusion (t_1) and the time of the end of metamorphic recrystallization (t_5). After Faure (1986).

the age since the rock cooled through the blocking temperature. Such a line is called a mineral isochron. It is also possible to plot isochrons based upon whole rock samples with different initial Rb:Sr ratios. These are called whole rock isochrons. A mineral isochron gives the time which has elapsed since rubidium and strontium were able to diffuse throughout a volume smaller than a hand specimen of the rock, while a whole rock isochron gives the time since Rb and Sr could be exchanged over several metres or more, which in an igneous intrusion is likely to have been when the magma was still liquid. For a suite of rocks with a simple cooling history (e.g. a set of granites from one intrusive complex) mineral and whole rock isochrons give the same age. But in regional metamorphic rocks from slowly cooled and re-heated areas such as the Adirondack Mountains of New York State, USA, whole rock isochrons may give the age of a period of igneous intrusion, while mineral isochrons give an age of later cooling. Metamorphism rarely makes the Rb:Sr ratios the same through large volumes of rock, but frequently does so in individual rock samples. The mineral and rock ages can be put together (Fig. 1.11) to determine:

(1) the time of the *end* of metamorphism, when the rock cooled through the blocking temperature
(2) the time of intrusion
(3) the initial $^{87}Sr : {}^{86}Sr$ ratio, which is important in igneous petrology.

This is merely an illustration of the ways in which isotope analysis can yield a great deal of information about the time and nature of metamorphic events. Much more information is to be found in Faure (1986) and Cliff (1985).

Electron-probe microanalysis

Another laboratory technique, **electron-probe microanalysis**, should be mentioned at this early stage, because its results are of direct value in interpreting metamorphic rocks seen under the petrological microscope. This technique determines the chemical compositions of small volumes at the surface of a polished mineral preparation. The polished mineral can be in a thin section (without a cover slip), so that the chemical compositions of minerals whose optical properties and textural relationships can be studied in the ways discussed in this book are obtained directly. Electron-probe microanalysers used for Earth Science research usually have a petrological microscope built in. They are widely used in modern studies of metamorphic rocks, and yield remarkable results in terms of the history of metamorphism of the rocks concerned.

The apparatus for performing electron-probe microanalysis is expensive, and so instruments are confined to laboratories in research establishments or universities. However, modern instruments are sufficiently simple and reliable to be used extensively by graduate students and frequently by undergraduates on advanced courses in mineralogy and petrology. They produce mineral analyses virtually instantaneously. It is worth saying something about their capabilities at this early stage in the book.

The microanalyser focuses a very narrow beam of energetic electrons onto the surface of the thin section (Fig. 1.12). The beam strikes the surface over a circular area about 1 μm in diameter, and the electrons excite the atoms in the mineral to emit X-rays. The wavelengths and energies of the X-ray photons emitted are characteristic of the different elements present in the mineral, and their intensities are approximately proportional to the concentrations of the elements.

The X-rays characteristic of each element must be distinguished, and this may be done in one of two ways. The X-rays may be discriminated on the basis of their characteristic energies (energy dispersion) or their characteristic

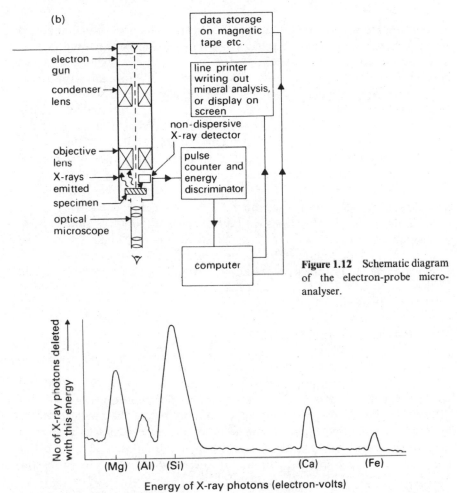

(b)

data storage on magnetic tape etc.

electron gun

condenser lens

line printer writing out mineral analysis, or display on screen

non-dispersive X-ray detector

objective lens

X-rays emitted

specimen

optical microscope

pulse counter and energy discriminator

computer

Figure 1.12 Schematic diagram of the electron-probe micro-analyser.

Figure 1.13 Energy-dispersion spectrum from the electron-probe microanalyser. The energy values of the peaks indicate which element produced the X-rays, and the heights of the peaks give the relative concentrations of the elements.

wavelengths (wavelength dispersion). Energy dispersion is the more rapid method and is more often used, but it is slightly less precise than wavelength dispersion, which is therefore still used occasionally. Only energy dispersion will be described here, and for more details of the apparatus and its use, the reader is referred to Potts (1987).

The X-rays emitted from the surface of the specimen are detected by a crystal which produces electric pulses whose energies are proportional to those of the X-ray photons. The electric pulses are then sorted and counted electronically, and the results presented as an energy-intensity spectrum (Fig. 1.13). The energy values of the peaks of such a spectrum are characteristic of the elements concerned, and their heights are approximately proportional to the concentrations of the elements. The spectrum is analysed by a microcomputer, and converted into a mineral analysis, making various assumptions about the emission process in the mineral, which need not concern us here. The result is usually printed out as an analysis in oxide weight percentages and as a chemical formula (see Chapter 2).

The electron-probe microanalyser can perform mineral analyses with a comparable accuracy to most other chemical and physical analytical techniques. It can analyse many spots in one mineral grain, and thus reveal variations in the composition of single crystals, which is known as **mineral zoning**. It can analyse small grains, down to < 5 nm.

It is possible to analyse even smaller grains by using an energy dispersion analysis system attached to a scanning electron microscope (**SEM**). There is a loss in the accuracy of analysis with this instrument, but it is good enough to distinguish between different mineral species in very small grains. The potentialities of the combined SEM and microprobe are still being explored by researchers, and promise to extend understanding of metamorphism still further.

Why study metamorphic rocks?

One aim in studying metamorphic rocks is to estimate the conditions of metamorphism, temperature, pressure, tectonic stress, presence of fluid and so forth. Such estimates give geologists an idea of conditions in the past at different depths in the crust and upper mantle, below the greatest depths which can be reached in mines and boreholes. Recent years have seen an increasing consistency in the results of such estimates, and also better consistency with geophysical estimates of present-day conditions within the Earth.

In arriving at such estimates, geologists studying metamorphic rocks are at a disadvantage compared with their colleagues working with igneous and sedimentary rocks. Processes of formation of igneous and sedimentary rocks can be observed occurring at the present day in various parts of the Earth's surface, and the **principle of uniformity** applied in the interpretation of similar rocks formed in the geological past. The metamorphic geologist has to work entirely upon a theoretical or experimental basis to interpret metamorphic rocks. In spite of this difficulty, the study of metamorphic rocks makes a major contribution to the understanding of processes of formation of crust and mantle, and the days are over when a geologist could leave the study of metamorphism to specialists.

Metamorphic rocks are also a significant source of economic raw materials, especially metal ores. Although the ores, like metamorphic rocks, usually have an ultimate igneous or sedimentary origin, their metamorphic history must be understood for the successful discovery of new deposits and effective mining methods when they have been found. It is also becoming clear that there is a close similarity between the processes involved in the low temperature metamorphism of sedimentary rocks and those responsible for the maturation and primary migration of oil and natural gas in sediments. The alteration of buried vegetable matter to form coal is also a process very similar to low temperature metamorphism. A lively exchange of ideas is going on, which is likely to illuminate both our understanding of metamorphism and the future exploration for oil, coal and natural gas deposits.

2 Rock and mineral compositions, and their relationship

This chapter introduces methods of study and description of metamorphic rocks and is intended for readers who are using a petrological microscope to study thin sections. It also discusses the relationship between rock analyses and mineral assemblages, and explains how these may be used to determine the conditions of metamorphism. Finally, a short outline is given of the principles of tectono-thermal modelling, which may be used to link the study of metamorphic rocks to plate tectonic processes.

Classification of metamorphic rocks

The broad classification of metamorphic rocks into contact metamorphic rocks, dynamic metamorphic rocks and regional metamorphic rocks has already been introduced, and is based upon the field relationships of the rocks concerned.

The character of any metamorphic rock will depend upon both its original sedimentary or igneous mode of formation, and upon its subsequent metamorphism. Methods of identification and description of metamorphic rocks in the field have to concentrate upon relationships between metamorphic rock units, and mineralogical and textural features which can be seen and described with the naked eye, perhaps aided by a hand lens. Such aspects of metamorphic rocks frequently are informative about the original sedimentary or igneous processes, but understanding of the metamorphism demands study of thin sections under a microscope (Fig. 2.1).

There is no generally agreed descriptive classification of metamorphic rocks, nor are there agreed definitions of such common metamorphic rock types as **schist**, **gneiss** and **amphibolite**. This is confusing for students and is particularly troublesome in the field, where lack of suitable names can lead to failure to distinguish significantly different metamorphic rock formations. An international commission has prepared proposals for a recognized scheme for classifying and naming metamorphic rocks, similar to those which have been successful for igneous rocks. It will be a great advance if such a scheme becomes accepted, but for the moment this section aims to provide guidance rather than hard and fast rules. It is hoped that the terms

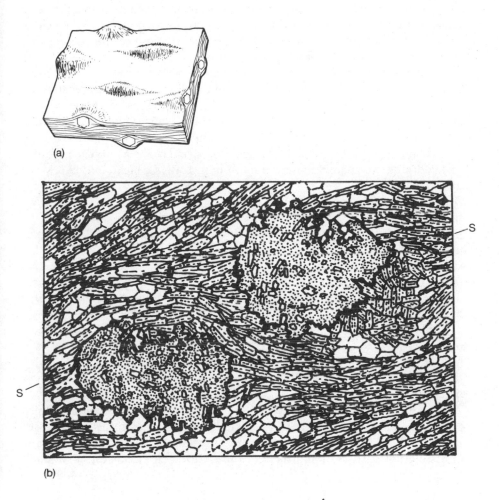

(a)

(b)

S

S

Figure 2.1 (a) Drawing of hand specimen of schist from Sulitjelma, Norway, with garnet prophyroblasts causing a change in the schistosity direction. (b) Thin section of the same schist, showing garnet porphyroblasts surrounded by a groundmass of biotite, muscovite, quartz and plagioclase feldspar. The schistosity is marked s–s. Scale bar 1 mm.

described here correspond reasonably well with the usage of British geologists, but in general readers are advised to try to understand and then follow locally used rock names. If there are no local terms, or if an area of metamorphic rocks is being described for the first time, then this chapter is intended to be helpful. The Geological Society of London's book 'The field description of metamorphic rocks' (Fry 1984) goes into more detail.

Primary sedimentary and igneous features

Many metamorphic rocks retain sufficient of their primary sedimentary or igneous features to be given sedimentary or igneous rock names in the field. Bedding is often preserved in sediments, and stratigraphical formations of rocks of different compositions and/or textures may be mapped. Where field study makes it certain that the units of metamorphic rock concerned are primary sedimentary formations, the conventional scheme for lithostratigraphical nomenclature should be employed (Hedberg 1976). **Formations**, **groups**, and **supergroups** may be defined and given appropriate geographical and rock-type names, e.g. Ardrishaig Phyllite Formation, Argyll Group, Dalradian Supergroup. Under the rules of stratigraphical nomenclature, metamorphic rock names may be used in lithostratigraphical unit names as alternatives to igneous or sedimentary rock names.

Like sediments, igneous rocks may retain sufficient of their primary features to be identified as lava flows, bedded pyroclastic rocks, minor intrusions or major intrusions. Lava flows and pyroclastic rocks may be named according to the lithostratigraphical conventions already mentioned, e.g. Tayvallich Lava Formation. Intrusive igneous rocks usually have rock names distinct from their volcanic equivalents and it may be appropriate to use these to distinguish intrusive from extrusive igneous rocks, even when both have undergone metamorphism.

If it is necessary to emphasize that a particular rock type has undergone metamorphism, this may be done in two alternative ways. The prefix 'meta' may be put before the igneous or sedimentary rock name, e.g. Tayvallich metabasalt, Knoxville metagraywacke, or a metamorphic rock name may be used in preference to an igneous or sedimentary one, e.g. Furulund schist. The context often makes it clear that metamorphic rocks are being described, making it tedious to keep repeating the prefix 'meta', and so unmodified sedimentary and igneous rock names are often used for metamorphosed or partly metamorphosed rocks, e.g. Loch Tay limestone, Sulitjelma gabbro. It is pedantic to insist that such widely used terms be modified merely to preserve a uniform convention for naming metamorphic rocks.

In many areas, however, primary igneous and sedimentary features have been completely obliterated by metamorphism. In other areas, it is not certain whether the boundaries between different lithological units represent sedimentary bedding or not. In such cases, the different metamorphic rock units should *always* be given metamorphic rock names, e.g. Man of War gneiss, Landewednack hornblende schist. The use of lithostratigraphical terms such as 'group' should be avoided. Several metamorphic rock units may be referred to collectively by the descriptive name '**complex**', e.g. The Lizard Compex (Hedberg 1976, p. 34). This is preferable to the older practice of adding the suffix 'ian' to a geographical name (i.e. 'the Malvern Complex' is preferable to 'Malvernian') because the use of the suffix leads to confusion with chronostratigraphic stage and system names. There are however some names such as 'Lewisian' which are too well established to be replaced.

Another confusion has arisen recently over the word '**terrain**', which is the British spelling. Large areas of deformed rocks in orogenic belts are now recognised as fragments carried over ocean basins from distant sites of original deposition, and these are called 'allochthonous terranes', 'suspect terranes' or 'displaced terranes', with 'terrane' spelt in the American way (Jones *et al.* 1982, Barber 1985), whereas 'metamorphic terrain' has long been used simply to describe an area of regional metamorphic rocks. Many of the rocks of allochthonous terranes are metamorphic, and some terranes are entirely made up of metamorphic rocks. The word **terrane** will be used in this book with the American spelling to mean 'allochthonous terrane', and the older, general use of 'terrain' will be avoided.

Mineral assemblages

Wherever possible, the minerals in a metamorphic rock should be identified. This should be done in the field if they are coarse-grained enough, and in all but the finest-grained metamorphic rocks all transparent minerals should be identified under the petrological microscope. They should be listed in decreasing order of abundance, e.g. biotite, muscovite, garnet, quartz, opaque minerals. Such a list is not necessarily a mineral assemblage list, because some minerals may be too fine-grained to identify, and also because the minerals may not have co-existed in equilibrium during metamorphism. Listing metamorphic minerals in the field is often very useful for recognizing changes in rock composition and **metamorphic grade**, saving later time and effort by noting significant changes, rather than returning to the laboratory and having to make another collecting trip.

The techniques for identifying minerals in the field are given in elementary textbooks of mineralogy (e.g. Gribble 1988). In studying metamorphic rocks, they should be applied to *small* grains, and students will find it especially helpful to study hand specimens in the lab, when they have the thin sections available for comparison. A useful way to teach yourself is to attempt the description of the hand specimen first, and then check how well you have done it by looking at the thin section. It is possible to do much better at identifying the minerals in metamorphic rocks in a hand specimen (and therefore in the field) than most beginners realise.

Metamorphic fabrics

Textural features of metamorphic rocks play a key role in their description and identification. Preferred orientation fabrics, such a **foliation** and lineation, are especially important. For example, the foliation fabrics **cleavage**

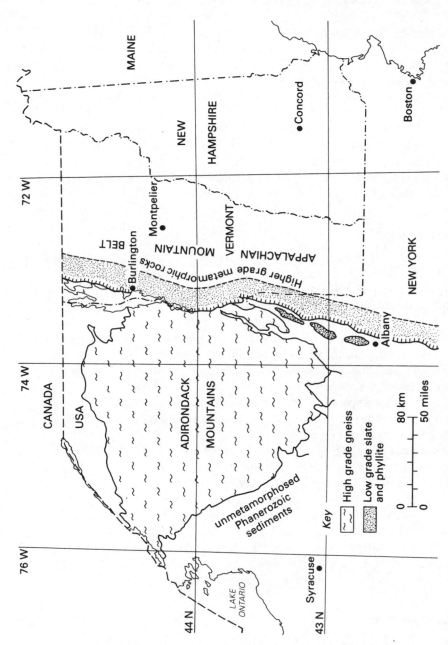

Figure 2.2 Geological map of the northeastern USA, showing high-grade Precambrian gneisses of the Adirondack Mountains, New York State, contrasted with low-grade slates and phyllites of the adjacent flank of the Appalachian orogenic belt.

and **schistosity** are diagnostic of slate and schist respectively. Another important type of foliation fabric is a segregation into light- and dark-coloured
layers, which is known as **gneissose banding**, and is diagnostic of most kinds
of gneiss.

The most important aspect of the fabric of metamorphic rocks is grain
size. Grain boundary regions of crystals have higher chemical potential
energies than the interiors, and since small crystals have a higher ratio of
surface area to volume than large crystals, grain sizes tend to increase during
metamorphism. The higher the temperature achieved during metamorphism,
the larger the grain size. Grain size also increases with the length of time the
rocks remained hot. Both processes are well illustrated by comparing the
rocks of the Adirondack Mountains of New York, USA, with those of the
nearby parts of the Appalachian orogenic belt in Vermont (Fig. 2.2). Both
suites of rock have undergone regional metamorphism, but the Adirondack
rocks were metamorphosed at higher temperatures, deeper in the crust, and
for a longer period than the Appalachian rocks. As a result, they are much
coarser-grained.

For metamorphic rocks from one metamorphic complex, which have
undergone metamorphism for approximately the same length of time, the
grain size increases with temperature, so that fine-grained rocks have probably been metamorphosed at low temperatures, coarse-grained ones at
higher temperatures. This is the basis for the rough-and-ready measure of
degree of metamorphism known as metamorphic grade. Fine-grained rocks,
metamorphosed at low temperatures, are said to be 'low grade metamorphic
rocks', coarse-grained rocks, metamorphosed at high temperatures, are said
to be 'high grade metamorphic rocks'. In Figure 2.2, the rocks of the
Adirondack Mountains are described as 'high grade gneisses', those of the
Appalachians as 'low grade phyllites'.

Metamorphic rocks often display different grain sizes in minerals formed
during metamorphism, and may also inherit variable grain sizes from their

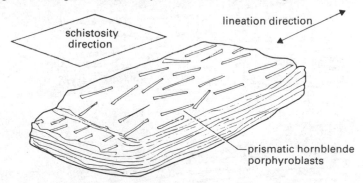

Figure 2.3 Hand specimen of schist containing porphyroblasts of prismatic hornblende, from
Furulund, Sulitjelma, Norway. Their random orientation within the schistosity planes is known
as '**garben texture**'.

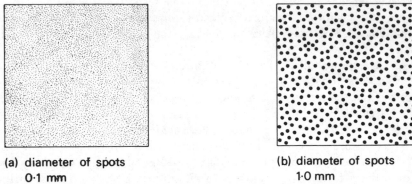

(a) diameter of spots
    0·1 mm

(b) diameter of spots
    1·0 mm

Figure 2.4   Grain size comparison digrams. (a) Dots 0.1 mm (b) Dots 1.0 mm.

igneous or sedimentary precursors. Rocks with large metamorphic crystals set in a finer-grained matrix are said to be porphyroblastic, and the large crystals are called porphyroblasts (Fig. 2.3). In such rocks, it is the *matrix* grain size which is indicative of metamorphic grade.

For the purposes of description and classification, it is useful to divide metamorphic rocks into coarse-grained, medium-grained and fine-grained, based upon matrix grain size. This is a less detailed division than that used in sedimentary rocks, but works in a similar way. Coarse-grained rocks have crystals more than 1 mm in diameter, medium-grained rocks have grains from 1 mm down to 0.1 mm, and fine-grained rocks less than 0.1 mm. The 0.1 mm limit is approximately the smallest size of grain which can be distinguished with a low-powered hand lens, and it is easy to recognize a crystal 1 mm in diameter after a little practice. Figure 2.4 displays patterns of dots, of 0.1 and 1 mm diameter, to help the reader become familiar with the grain

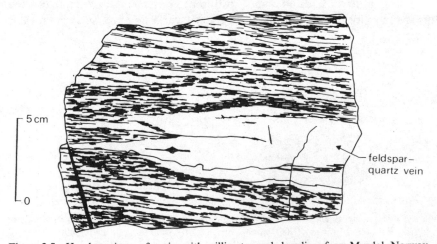

feldspar–
quartz vein

5 cm

0

Figure 2.5   Hand specimen of gneiss with millimetre-scale banding, from Myrdal, Norway.

Table 2.1 Scheme of classification of metamorphic rocks in the field.

Primary rock composition	Main mineral groups present	Fine grained < 0.1 mm	Medium grained 0.1 mm– 1.0 mm	Coarse grained > 1 mm homogeneous	banded
shale 'pelitic rocks'	phyllosilicates quartz	slate	phyllite	schist	gneiss
sandstone 'psammitic rocks'	quartz	quartzite	quartzite→	quartzite	banded quartzite
sandy shale 'semi-pelite'	quartz phyllosilicates	sandy slate	semi-pelitic phyllite	semi-pelitic schist	gneiss
limestone	calcite dolomite	marble	marble	marble	banded marble
marl	phyllosilicates calcite	calc–slate	calc–phyllite	calc–schist	calcareous gneiss
rhyolite, granite	K-feldspar, quartz phyllosilicates	halle-flinta	granitic gneiss	granitic gneiss	granitic gneiss
basalt, gabbro	plagioclase feldspar amphiboles		greenschist greenstone	amphibolite	hornblende gneiss pyroxene gneiss
pyroxenite peridotite dunite	serpentine talc		serpentinite talc schist soapstone		ultrabasic gneiss foliated u-basics

sizes. Alternatively, published grain-size scales for sedimentary rocks may be used. These give the 1 mm size (quoting it as 1000 μm), but then progressively halve the millimetre scale, illustrating grain sizes of 0.5, 0.25 and 0.125 mm (i.e. 125 μm), rather than the 0.1 mm taken in Figure 2.4 to be the boundary between medium and fine grain sizes. For the rough and ready estimation we are discussing, this small difference doesn't matter.

Coarse-grained metamorphic rocks frequently display compositional banding or layering (Fig. 2.5). This can be shown to be a result of metamorphism, rather than the survival of sedimentary layering. In Table 2.1, coarse-grained metamorphic rocks with layering are distinguished from coarse-grained rocks without layering.

Rock compositions

Table 2.1 gives a rough classification of common metamorphic rocks, based upon the grain-size scale just introduced, and on a crude subdivision of original rock compositions. The 'bulk' chemical compositions of rocks are of the most interest for metamorphic geologists, because the original minerals, grain sizes and textures become obliterated by metamorphism.

In the table sedimentary rocks are divided into compositional groups characterized by shale (or mudrock), quartz-rich sand, sandy shale, limestone (or carbonate rock), and limey shale (or marl). Igneous rocks are divided into acid, basic and ultrabasic categories.

Chemical compositions of metamorphic rocks

The bulk chemical compositions of rocks are expressed quantitatively in the form of chemical analyses. Because oxygen is the most abundant element in the lithosphere, it is conventional to express the compositions of rocks as lists of oxides, expressed as weight percentages, for major elements. Trace elements are expressed as parts per million (p.p.m.) by weight, or even as parts per billion (p.p.b.).

Ten major oxides are usually quoted, and make up as much as 99% of most silicate rocks. In the usual order in which they are given they are SiO_2, Al_2O_3, Fe_2O_3, FeO, MgO, CaO, Na_2O, K_2O, 'H_2O^+' and 'H_2O^-'. The separate values quoted for Fe^{+2} and Fe^{+3} oxides indicate that in most rocks the iron is not completely oxidized. The expressions 'H_2O^+' and 'H_2O^-' refer to water driven off from the rock during analysis of a crushed powder. Experience has shown that the water adhering to or adsorbed upon mineral grains, and that loosely held in cracks, is reliably determined by measuring the weight loss when the powder is held for several hours in an oven at a temperature of 105°C. This is quoted as 'H_2O^-'. There may also be water locked in the structure of minerals, and hydroxyl groups combined with their cations. This is driven off by much stronger heating, and is quoted as 'H_2O^+'. A number of minor metallic elements are usually quoted as oxide weight percentages, the exact list given depending upon the type of rock being analysed. Minor elements often quoted as oxides include TiO_2, Cr_2O_3, MnO, BaO, P_2O_5, CO_2 and SO_3. Some non-metals, such as fluorine and chlorine, present in the rocks as anions (F^-, Cl^-) are not quoted as oxides but in element weight percentages. Non-metals often quoted in this way include F, Cl and S. In some rocks any of these elements may be major components. Finally, the total is given of the weight percentages of the oxides and elements quoted. This is intended to indicate the reliability of the analysis, totals between 99.7% and 100.5% being regarded as acceptable.

This form of tabulation of rock analyses is not very useful for petrological studies of metamorphic rocks, which are concerned with the expression of rock composition in metamorphic mineral assemblages. It is the relative *numbers* of atoms of the different elements in the rocks which are of interest. Because of the oxidized condition of rocks, the relative numbers of molecules of different oxides in the analyses are often compared. To obtain these, the weight percentages of the oxides are divided by their molecular weights, of which a list is given in the Appendix. In order to make the distinction between weight percentages and molecular proportions clear, square brackets will be put round the oxide formula when the molecular proportion is under discussion (e.g. $[SiO_2]$, $[Fe_2O_3]$, $[TiO_2]$). When weight percentages are being referred to, the formulae or numbers will be left without brackets.

It is possible to recalculate the analyses of metamorphic rocks in terms of simplified mineral compositions (e.g. wollastonite $CaSiO_3$, albite plagioclase $NaAlSi_3O_8$). The result of such a calculation is a metamorphic rock norm. Norms are usually calculated by following one of the standard schemes for norm calculations on igneous rocks (Cox *et al.* 1979), or using a computer program based on them. The calculations give a list of minerals, and their relative proportions, which may be applied either to estimate the original composition of the **protolith**, especially if it was an igneous rock, or to consider possible metamorphic mineral assemblages, which have subsequently broken down.

Mineral compositions

Because minerals are chemical compounds, the relative proportions of elements including oxygen, and therefore of oxides in their analyses are usually fixed. For example $[CaO]:[SiO_2]$ in diopside ($CaMgSi_2O_6$) is $1:2$, in tremolite ($Ca_2Mg_5Si_8O_{22}(OH)_2$) $1:4$. The chemical formula of diopside may alternatively be written $[CaO].[SiO_2]$, and that of tremolite $[2CaO].[5MgO].[8SiO_2].[H_2O]$, which makes the relationships clearer. Where one metallic ion may substitute for another in the same position in the crystal structure of a mineral, a phenomenon known as **diadochy** or **solid solution**, the ratio of the two may have any value (e.g. $[MgO:FeO]$ in diopside–hedenbergite) or may have any value within a fixed range (e.g. $[Al_2O_3]:[SiO_2]$ in plagioclase feldspars) for which the range is from $1:3$ in albite ($NaAlSi_3O_8$) to $1:1$ in anorthite ($CaAl_2Si_2O_8$). The molecular ratios of the sums of the substituting oxides remain fixed relative to other oxides in the mineral. For example $[FeO + MgO]:[SiO_2]$ in diopside–hedenbergite, $Ca(Mg, Fe^{+2})Si_2O_6$, is $1:2$.

A mineral such as a clinopyroxene of the diopside–hedenbergite series may be a member of an equilibrium mineral assemblage in a metamorphic rock. As the next section will show, this may be expressed by saying that under the conditions of metamorphism, it was a single chemical **phase**. The chemical composition of the mineral may vary in a regular way between the two end-member compositions, and be expressed as the molecular percentages of these two (e.g. 68% $CaMgSi_2O_6$, 32% $CaFe^{+2}Si_2O_6$) or more briefly $Di_{68}Hd_{32}$.

Much of this is probably already familiar from your knowledge of mineralogy, but it is repeated here because it is crucial to the following section.

The Phase Rule

Goldschmidt (1911) examined hornfelses of variable composition surrounding igneous intrusions in the Oslo region of Norway, and recognized that there is a simple relationship between rock compositions and mineral assemblages in metamorphic rocks which reached a state of thermodynamic equilibrium during metamorphism. The relationship is the expression in rocks of a simple law of physical chemistry, the **Phase Rule**. The Phase Rule is expressed by the equation:

$$P + F = C + 2 \qquad \qquad \text{(i)}$$

P, F and C are simple whole numbers (integers). P is the number of phases in the **system**, that is to say the number of physically distinct kinds of substance which can be distinguished. F is the number of modes of variation or **degrees of freedom** (also known as the **variance**) of the system. C is the number of chemical **components** in the system. The rule applies to any chemical system which has attained a state of thermodynamic equilibrium. It may be applied either to a rock undergoing metamorphism or to an experimental system under observation in the laboratory.

Figure 2.6 illustrates the meaning of the Phase Rule by considering a series of simple experiments in a sealed vessel, containing pure H_2O only.

The temperature in the vessel may be changed by heating or cooling it from outside, and the pressure by means of a piston at one end. Figure 2.6 illustrates the state of the experimental system (everything inside the vessel) under different conditions during the experiments. Three different phases appear, liquid water, ice and water vapour (steam). All have the same composition, H_2O, and H_2O is the only substance inside the vessel during the experiments. At all stages in the experiments, therefore, there is only one chemical component to be considered, H_2O; and thus C in the Phase Rule equation (i) is always 1. The values of P, F and C in the equation during each experiment are given, under the temperature and pressure, to the right of the illustrations in Figure 2.6.

Figure 2.6a shows the system at a temperature of 30°C and under a pressure of 100 kPa (1 bar, approximately atmospheric pressure). Under these conditions the system contains only one phase, liquid water. Therefore P = 1. It follows from equation (i) that F = 2, i.e. the system has two degrees

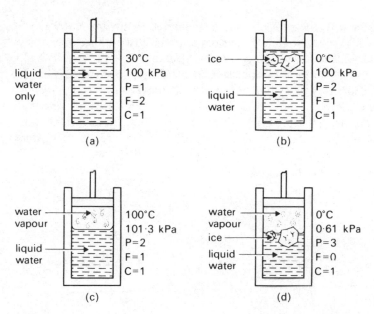

Figure 2.6 Experiments in the one-component system H_2O. For a full description, see the text.

of freedom. What does this mean? It is possible to change *either* the temperature, *or* the pressure from their values shown in Figure 2.6a by a small amount without changing the phases present in the system. If the temperature is increased to 31°C or decreased to 29°C, the vessel will still only contain liquid water. If the pressure is increased to 110 kPa or decreased to 90 kPa, the vessel will also only contain liquid water. The temperature and pressure may be varied independently. The system is said to be **divariant**.

In Figure 2.6b, the system contains two phases, liquid water and solid ice. $P = 2$ and so from equation (i) $F = 1$. The system is said to be **univariant**. In this case a small change in either temperature or pressure would cause a fundamental change in the system, reducing the number of phases present from two to one. An increase of temperature to 1°C, keeping the pressure constant at 100 kPa, would cause all the ice to melt. A decrease of temperature to − 1°C would cause all the water to freeze. An increase of pressure to 110 kPa, if temperature is kept exactly constant at 0°C, would cause the ice to melt, a decrease of pressure to 90 kPa would cause all the water to freeze. Only if the *ratio* of temperature to pressure were kept constant would the system continue to contain two phases, ice and water.

In Figure 2.6c, the system also contains two phases, water and water vapour (or steam). In this case an increase of temperature to 101°C will cause all the water to change to steam, if the pressure is kept at 101.3 kPa, while a decrease in temperature to 99°C will cause all the steam to condense to water. An increase in pressure to 102.3 kPa will cause all the steam to condense to water, if the temperature is kept at 100°C, and a decrease in pressure to

100.3 kPa will cause all the water to change to steam. At 99°C water and steam will coexist, so that the system remains a two-phase one, at a pressure of 97.9 kPa, and at 101°C a two-phase system will be preserved at 105.0 kPa. For any temperature, the pressure is fixed, and *vice versa*. This is strictly what is meant by describing the system as univariant.

Figure 2.7 shows the relationships between water, ice and water vapour on a temperature–pressure graph. It can be seen that the divariant states of the system, when there is only one phase present, are represented by areas on the

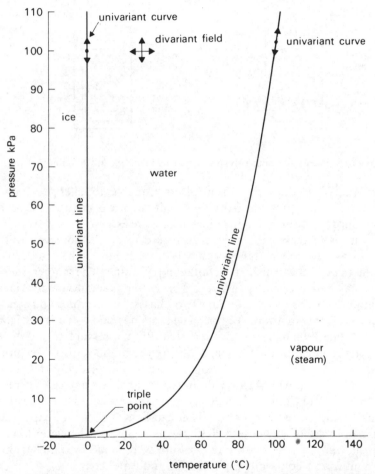

Figure 2.7 Phase diagram for the one-component system H_2O from − 20°C to 140°C and from 0 kPa to 140 kPa. The pressure at which ice and water co-exist in equilibrium falls very rapidly with increasing temperature, by about 10 000 kPa for every 1°C, and thus the univariant curve for ice and water is effectively isothermal on this diagram. The univariant state of the system shown in Figure 2.6 (b) can therefore be discussed (as in the text) as one in which a change in temperature alters the system from a two-phase condition into a one-phase condition, while a change in pressure has no effect.

graph. The univariant temperatures and pressures are represented by curves separating the divariant fields, and the particular examples of uni-variant conditions in Figures 2.6b and 2.6c are indicated. Lastly, there is an **invariant triple point** at which ice, water and vapour all coexist in equilibrium. The temperature and pressure for this state of the system are unique, and any change in them will cause one or two phases to disappear. This state of the system is illustrated in Figure 2.6d. Figure 2.7 is called a **phase diagram**. It is a simple diagram because the system described is a simple one, with only one component. One component systems do exist in natural rocks, one example being the system composed of SiO_2.

But as we have already seen, most metamorphic rocks have several components and contain several different mineral phases. From equation (i) we can see that as the number of components increases, the number of phases also increases. If the rock had attained a state of equilibrium during metamorphism, the minerals in the rock can be regarded as phases and the Phase Rule equation (i) applies. Thus the Phase Rule links the number of minerals in an equilibrium metamorphic assemblage list with the number of chemical components in the rock analysis. This explains the fact that the number of different minerals present in most metamorphic rocks is small, usually less than five or six.

How many degrees of freedom are likely to be present in rocks undergoing metamorphism? In the first chapter, we saw that thermal modelling suggests that temperature and pressure during metamorphism are likely to be determined by the rate of heat flow through the lithosphere, i.e. they are external to the rock undergoing metamorphism, similar to the externally applied temperature and pressure illustrated in Figure 2.6. So for rocks undergoing metamorphism, the number of degrees of freedom must be at least two. Equation (i) can thus be modified for metamorphic rocks to be:

$$P < C \qquad \qquad (ii)$$

P is now simply the number of metamorphic minerals in the assemblage list, and C the number of chemical components in the rock system. C can be determined by finding out how many oxide ratios are needed to describe the compositions of all the metamorphic minerals, and is smaller than the number of major oxides listed in a chemical analysis. This relationship was first demonstrated by Goldschmidt in the Oslo hornfelses, and is known as **Goldschmidt's Mineralogical Phase Rule**. Most modern interpretations of metamorphic rocks are based upon the application of this rule, or to be more accurate, a modification of it to allow for the presence of a vapour phase during metamorphism in most cases. It is central to the recognition of **metamorphic facies**, which is a widely-used method of summarizing the conditions of metamorphism of large suites of metamorphic rocks.

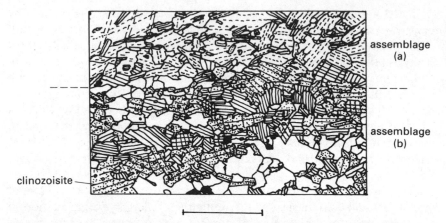

Figure 2.8 Contrasting equilibrium mineral assemblages in skarn (metamorphosed carbonate rock) from Connemara, Ireland. Assemblage (a) hornblende + calcite + clinozoisite, assemblage (b) clinozoisite + diopside + quartz + opaques. The rock therefore shows two separate domains of equilibrium, with different mineral assemblages. Scale bar 1 mm.

Recognition of equilibrium in metamorphic rocks

Although metamorphic rocks frequently show a simple mineral assemblage list, reflecting the attainment of thermodynamic equilibrium and the Phase Rule, this equilibrium is only attained for part of the metamorphic history of the rock, as displayed in its P–T–t path. Consideration of metamorphic rocks themselves also shows that equilibrium is attained only over quite small volumes of the rock material. Metamorphic rocks are frequently inhomogeneous on a hand-specimen scale, with volumes less than 10 mm across having different chemical compositions and different metamorphic minerals. It is often possible to recognize different metamorphic mineral assemblages, reflecting different parent rock compositions, in a single thin section (Fig. 2.8). The volume over which the metamorphic minerals crystallized in equilibrium with one another (or close to equilibrium) is known as the **domain of equilibrium**. It may be quite large, as in some of the gneisses described in Chapter 8, or very small, as in some of the **slates** described in Chapter 5.

The determination of metamorphic temperatures and pressures depends upon recognition of the extent of attainment of equilibrium, and this is usually done by careful study of thin sections under the petrological microscope. It is helped considerably by electron microprobe analyses of individual mineral grains. Simple rules for recognising that a metamorphic rock has attained equilibrium cannot easily be formulated. It is a matter of experience, and still to some extent of individual opinion. To help you develop your own judgement, part of the aim of this book is to give information about a number of case histories of regions of metamorphic rocks, with discussion of the degree of attainment of equilibrium, or lack of it. Provided that the list of minerals in the rock appears to be related to the chemical composition in accordance with the Phase Rule, it is usual to regard the rock as having attained equilibrium, unless there are textural features strongly to the contrary. The list of minerals in the rock is called an equilibrium mineral assemblage, or mineral assemblage for short.

Hornfelses are massive metamorphic rocks, formed close to the contacts of igneous intrusions (Chapter 3). They have uniform grain sizes, and the grains themselves approach equant shapes (except for the mica flakes) (Fig. 3.3). This type of metamorphic rock frequently displays textural and thermodynamic equilibrium. The characteristic fabric is called **granoblastic texture**, because it gives the hornfels an even-grained, granular appearance when it is studied under a hand lens. It is the usual texture of hornfelses, but also occurs in many regional metamorphic rocks. It arises by processes of growth of crystals in the rock when it was free from externally applied stress.

In all metamorphic rocks, crystals which are increasing in size compete for space, and come into contact at their crystal boundaries. Atoms in crystals

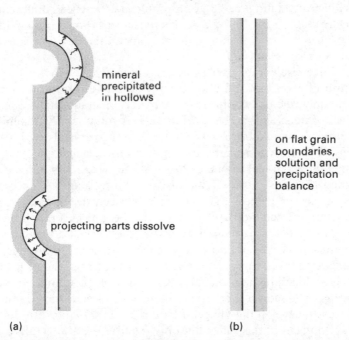

mineral
precipitated
in hollows

on flat grain
boundaries,
solution and
precipitation
balance

projecting parts dissolve

(a)   (b)

Figure 2.9 Selective solution and precipitation of crystals at grain boundaries tend to make them become flat. (a) Cross section through a boundary between two crystals of the same mineral. The 'peninsula' has a slightly higher chemical potential energy than the rest of the source, and therefore tends to dissolve, while the 'bay' has a slightly lower energy, and therefore tends to be a site of precipitation. (b) When these processes reach equilibrium, the surfaces have become flat and have the lowest possible chemical potential energy.

have a higher potential energy at and near the surface then they do in the interior, because of the irregular and broken chemical bonds between atoms which must exist at the surface. Therefore, crystals growing without the influence of stresses in the rock tend to have the minimum surface area possible, to enable the potential energy of the rock as a whole to be as low as possible. The geometrical shape with the lowest surface area for its volume is a sphere, but crystals in a rock do not adopt this shape, because as they grow they impinge on one another, and because in crystalline structures 'rational' faces (which can be given Miller index numbers) have lower surface energies than random sections through the crystal.

Where two crystals are alongside one another, the shape for the interface which has the lowest area, and thus the lowest surface energy, is a plane (Fig. 2.9). Where three crystals impinge, the lowest surface energy configuration is for three planar interfaces to intersect at 120° (Fig. 2.10), provided that all three crystals have similar surface energies (which will be the case if they are all of the same mineral, e.g. quartz). Thus crystals impinging on one another with a low surface energy configuration will have planar interfaces with adjacent crystals, and edges where two faces meet at 120°. In the hornfels illustrated in Figure 3.3, for example, it can be seen that many **triple junctions** are close to the 120° shape. There is a variation in apparent angles because of the random orientation of the plane of the section relative to the edges where three crystals join.

A useful exercise can be carried out to emphasize this point, studying a thin section of a rock with granoblastic texture and a mineral assemblage containing only one or two minerals. A massive metaquartzite or a mono-minerallic hornfels (e.g. **saccharoidal marble**) is suitable. The magnification of the microcope should be selected so that 100 or so crystals are visible, and the angles of intersection at the triple junctions measured by making the intersecting faces parallel to one or other of the cross-hairs in the eyepiece, and reading the value on the scale of the rotating stage. A check on the accuracy of measurement is that the angle between the three faces at each triple junction should add up to 360°. If the angles are plotted on a histogram, they should form an even distribution about a median value of 120°. An alternative way of doing the exercise is to take a photomicrograph of the thin section, and carry out the same measurements using a protractor.

The shape which preserves planar faces, and triple junctions close to 120°, is a truncated octahedron (Fig. 2.11). There is a more detailed discussion of the tendency to develop this shape in Hobbs *et al.* 1976, pp. 110–13.

The development of this shape not only requires the grains to grow in the absence of tectonic stress, it also only applies to grains which are isotropic, or close to isotropic. For the processes of growth of mineral crystals in the majority of metamorphic rocks, framework silicates such as quartz and feldspars behave as though they are isotropic, so that quartzites develop the **equilibrium fabric** similar to that shown in Figure 3.3, in many contact

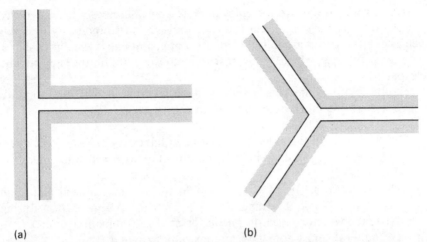

(a)   (b)

Figure 2.10 Selective solution and precipitation of crystals of the same mineral tends to make them develop 120° triple junctions. (a) The 90° angles of the two crystals on the right have a higher chemical potential energy than the rest of the crystal faces, and therefore tend to dissolve. (b) When equilibrium is attained, there are three 120° angles, which have a lower total potential energy than the arrangement in (a).

Figure 2.11 The truncated octahedron crystal shape has all interfacial angles close to 120°, and therefore tends to be widely developed in rocks which have attained a state of textural equilibrium.

sieve textured
hornblende

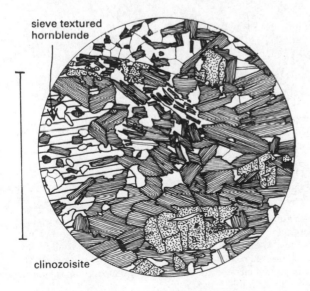

clinozoisite

Figure 2.12 Schist with the equilibrium mineral assemblage biotite + clinozoisite + hornblende + plagioclase + quartz + opaques. Skaiti Supergroup, Sultijelma, Norway. Scale bar 1 mm.

aureoles. More strongly anisotropic minerals, such as amphiboles and micas, develop anisotropic shapes, especially if they are surrounded by minerals such as quartz, so that amphiboles are prismatic, and micas tabular in shape, even in hornfelses which show textural equilibrium in the quartz and feldspar.

Textural relationships which must be demonstrated if the list of minerals in a rock are to be taken as an equilibrium assemblage are:

(1) Each mineral in the assemblage list must have a boundary somewhere within the proposed domain of equilibrium with all other members (Fig. 2.12).
(2) The texture must be of a type thought to have crystallized mainly in response to a rise in temperature, rather than fragmentation during dynamic metamorphism or igneous crystallization from magma.
(3) The minerals must not show compositional zoning.
(4) There must not be obvious replacement textures such as reaction rims round minerals, or alteration along cracks.

Figure 2.12 shows a regional metamorphic rock which has an equilibrium mineral assemblage, while Figure 2.13 depicts one with more minerals in it

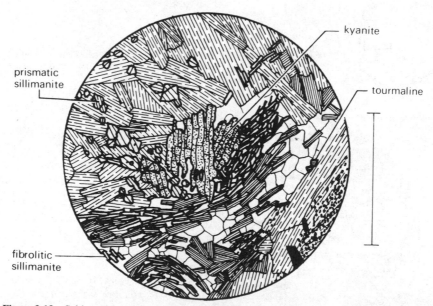

prismatic
sillimanite

kyanite

tourmaline

fibrolitic
sillimanite

Figure 2.13 Schist with a disequilibrium mineral list, biotite, quartz, sillimanite, kyanite, opaques, tourmaline. Skaiti Supergroup, Sulitjelma, Norway. Kyanite and sillimanite are both members of the one-component system Al_2SiO_5, and can therefore only co-exist in equilibrium if the temperature and pressure conditions of metamorphism fall exactly onto the kyanite/sillimanite univariant curve. The P–T–t path of metamorphism in the Skaiti Supergroup is known (Fig. 7.21), and crosses this curve rather than following it.

than the Phase Rule would predict, and textures strongly indicating disequilibrium. In the rock of Figure 2.12, the grain size is relatively uniform, grain boundaries are planar, and the faces at triple junctions meet at 120°. This indicates that the rock is close to a state of textural equilibrium, which suggests that it was also close to a state of thermodynamic equilibrium during metamorphism. By contrast, the crystals in the rock in Figure 2.13 vary considerably in size, some minerals form rims around others, and grain boundaries are not flat, or in definite crystallographic directions. The rock displays marked textural disequilibrium, matching its thermodynamic disequilibrium.

It may still be possible, where minerals are compositionally zoned, or reaction rims are present, to work out conditions of metamorphism by considering equilibrium in domains which are smaller than an individual grain.

When we are satisfied that the list of minerals present in a rock constitute an equilibrium mineral assemblage, in this book the minerals will be separated by plus signs, thus:

diopside + anorthite + grossular garnet + quartz

Composition assemblage diagrams

One very useful application of the Phase Rule to metamorphic rocks is that it permits the preparation of diagrams, which represent the mineral assemblages of a range of different rock compositions. The diagrams rest upon similarities between the geometrical properties of solid or planar figures, and the relationships between components and phases in the Phase Rule. For example, if a system is composed of three components, all possible relative proportions of those components can be displayed in an equilateral triangle.

A problem in preparing composition–assemblage diagrams is that the number of major components in metamorphic rocks is greater than three, while a two-dimensional planar diagram can only represent phase relationships completely for a system of three components or less. This problem is overcome by making projections from the composition of one phase which is present in all the mineral assemblages under consideration. Diagrams are usually designed to display changing phase relationships in rocks with a restricted range of composition, such as Na-poor pelitic rocks, or basic igneous rocks. In this book, no attempt will be made to present composition–assemblage diagrams applicable to all metamorphic rocks. A variety of special diagrams will be used, of which two, the ACF diagram and the AFM diagram, will be explained further in this chapter. Other types of diagram will be explained when they are presented in later chapters.

The ACF diagram

ACF diagrams are used to represent the compositions of basic igneous rocks, and their accompanying mineral assemblages. They are also applicable to rocks containing calcium and magnesium carbonates and silicate minerals (i.e. 'marls'). Five major components are taken to make up the rock: SiO_2, Al_2O_3, FeO, MgO and CaO. FeO and MgO are assumed to be able to substitute for one another in all proportions in all the minerals which are plotted on the diagrams. It is assumed that all the rocks plotted contain SiO_2 in excess, so that quartz is a member of all mineral assemblages. This leaves three components for plotting: $[Al_2O_3]$, $[CaO]$ and $[FeO] + [MgO]$. These components are the A, C and F of the title of the diagram. There are corrections applied to incorporate minor elements. $[FeO]$ is reduced to allow for $[TiO_2]$ and $[Fe_2O_3]$ in ilmenite and magnetite, respectively, and $[CaO]$ is corrected to allow for combination with $[P_2O_5]$ to form apatite. The chemical formula of apatite is $Ca_5(PO_4)_3(OH, F, Cl)$, which may be written $10CaO.3P_2O_5.OH, F, Cl$, so $10/3 \times [P_2O_5]$ is subtracted from the $[CaO]$ total. $[Na_2O]$ and $[K_2O]$ are assumed to be present in plagioclase feldspar, $[K_2O]$ as the small amount which substitutes for $[Na_2O]$ in albite. Since the formula of albite may be written $Na_2O.Al_2O_3.6SiO_2$, and that of orthoclase $K_2O.Al_2O_3.6SiO_2$, $[Na_2O]$ and $[K_2O]$ are subtracted from the $[Al_2O_3]$ total, assuming that plagioclase is always present in the assemblage list.

Various corrections are applied for minor components, which can be best understood by considering a worked example, in Table 2.3. The calculation is easily performed using a pocket electronic calculator. The simplest type of 'four function' calculator will do, but calculations can be speeded up using a memory function. Readers with a programmable calculator will find it interesting to work out the necessary program, and the calculation can easily be adapted for a home computer. The procedure to be followed during the calculation, beginning from the rock analysis, is as follows:

(1)   Divide weight percentages (Column 1) by molecular weights (Column 2) to obtain molecular proportions (Column 3).

(2)   Correct $[Al_2O_3]$ by subtracting the aluminium present in the feldspars:

$$A = [Al_2O_3] - [Na_2O] - [K_2O]$$

(3)   Correct CaO by subtracting the calcium present in apatite $(10CaO.3P_2O_5)$ and calcite $(CaO.CO_2)$.

$$C = [CaO] - 10/3 \times [P_2O_5] - [CaO_2]$$

(4)   Correct $[FeO]$ by subtracting the ferrous iron in ilmenite and magnetite:
$$[FeO] \text{ (corrected)} = [FeO] - [TiO_2] - [Fe_2O_3]$$

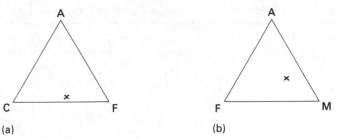

Figure 2.14 (a) ACF triangular composition–assemblage diagram for the calculation in Table 2.2 (b) AFM triangular composition–assemblage diagram for the calculation in Table 2.3.

$$F = [MgO] + [FeO] \text{ (corrected)}$$

(5) Determine the sum of A + C + F.
(6) Express A, C and F as percentages of (A + C + F). An example of the calculation in a metamorphosed basic igneous rock is laid out in Table 2.2, and the result is plotted on an ACF triangle in Figure 2.14a.

The AFM diagram

AFM diagrams are used to represent rock compositions and mineral assemblages in pelitic rocks. They were invented by Thompson (1957), and use a projection method to deal with the problem of having four major components in these rocks. Five major components of rock composition are considered initially: SiO_2, Al_2O_3, FeO, MgO and K_2O. The minor components Fe_2O_3, TiO_2 and P_2O_5 are allowed for by corrections to appropriate major components. Pelitic rocks usually contain a proportion of quartz in their mineral assemblages, indicating that SiO_2 is present in a large enough amount to form all the other silicate minerals in the assemblage, and still leave enough to crystallize as quartz. Therefore, provided that the caption to the diagram states that quartz is present in the rocks as well as the minerals displayed by the diagram, SiO_2 need not be considered further. This reduces the number of major components considered to four, and they form the corners of the tetrahedral diagram drawn in perspective in Figure 2.15. This is called an AKFM tetrahedron.

Because pelitic rocks almost always contain white mica, the number of components is reduced to three by projecting the rock composition which lies within the AKFM tetrahedron onto the base, from the composition of muscovite (Fig. 2.15). In the final diagram, therefore, three components are plotted; A (aluminium), F (iron) and M (magnesium), and the mineral assemblages always contain quartz and white mica (muscovite) in addition to the minerals shown on the diagram.

Table 2.2 Calculation of ACF values for an analysed basic igneous rock, plotted on Fig. 2.14a.

	1.	2.	3.
SiO_2	48.62	60.07	0.8094
TiO_2	1.84	79.89	0.0230
Al_2O_3	9.71	101.82	0.0954
Fe_2O_3	5.40	159.68	0.0338
FeO	4.45	71.84	0.0619
MnO	0.16	70.93	0.0023
MgO	7.69	40.31	0.1908
CaO	9.89	56.07	0.1764
Na_2O	3.94	61.97	0.0636
K_2O	0.31	94.20	0.0033
H_2O (total)	6.80	18.01	0.3776
P_2O_5	0.18	141.92	0.0013
CO_2	0.86	44.00	0.0195
Total	99.85		

$A = [Al_2O_3] - [Na_2O] - [K_2O] = 0.0954 - 0.0636 - 0.0033 = 0.0285$
$C = [CaO] - 10/3 \times [P_2O_5] - [CO_2] = 0.1764 - 0.0043 - 0.0195 = 0.1526$
$F = (MgO) + [FeO] - [Fe_2O_3] - [TiO_2] = 0.1908 + 0.0619 - 0.0338 - 0.0230 = 0.1959$
$A + C + F = 0.3770$
$A = 7.6\%$ $C = 40.4\%$ $F = 52.0\%$

*Column 1: weight percentages of oxides in a metamorphosed basalt from Mutki, Turkey, analysed by R. Hall.
Column 2: molecular weights from Appendix I.
Column 3: molecular proportions calculated as explained in text.
†Mineral assemblage list for this rock:
albite + chlorite + sodic amphibole + clinopyroxene + sphene + opaques

The calculation is performed as follows:

(1) Divide the weight percentages of the oxides in the analysis (Column 1) by their molecular weights (Column 2). A list of molecular weights of oxides and elements which might appear in rock or mineral analyses is to be found in the Appendix. The results of the divisions are in Column 3.

(2) [FeO] is corrected to allow for the presence of ferrous iron in ilmenite ($FeO.TiO_2$) amd magnetite ($FeO.Fe_2O_3$).

$$F = [FeO] - [TiO_2] - [Fe_2O_3]$$

(3) The projection from muscovite composition onto the base of the AKFM tetrahedron is performed as follows. The chemical formula of pure muscovite may be written in the form $K_2O.3Al_2O_3.6SiO_2.2H_2O$. Therefore three times the molecular proportion of K_2O is subtracted from the molecular proportion of Al_2O_3. The sodium present in the

rock is assumed to be present as albite plagioclase ($Na_2O . Al_2O_3 . 6-SiO_2$), and therefore $[Na_2O]$ is subtracted from the $[Al_2O_3]$ as well.

$$A = [Al_2O_3] - 3 \times [K_2O] - [Na_2O]$$

(4) $$M = [MgO]$$

(5) The totals for A, F and M are added.
(6) Determine the proportions of the total attributed to A, F and M, and express the results as percentages.

Table 2.3 shows a worked calculation for an analysed schist from Turkey, and the result is plotted in Figure 2.14b.

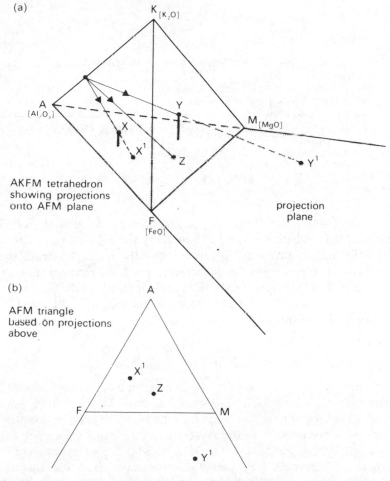

(a)

K [K₂O]

A [Al₂O₃]

Y

M [MgO]

X

X¹

Z

Y¹

AKFM tetrahedron
showing projections
onto AFM plane

projection
plane

F [FeO]

(b)   A

AFM triangle
based on projections
above

X¹

Z

F   M

Y¹

Figure 2.15   Projection of rock and mineral compositions onto the base of an AKFM tetrahedron to form a Thompson AFM diagram.

Table 2.3 Calculation of AFM values for a pelitic rock, plotted on Figure 2.14b.

	1.	2.	3.
SiO_2	63.62	60.07	1.0591
TiO_2	0.94	79.89	0.0118
Al_2O_3	15.81	101.82	0.1553
Fe_2O_3	3.09	159.68	0.0194
FeO	4.03	71.84	0.0561
MnO	0.08	70.93	0.0011
MgO	2.77	40.31	0.0687
CaO	1.76	56.07	0.0314
Na_2O	2.43	61.97	0.0392
K_2O	2.80	94.20	0.0297
P_2O_5	0.20	141.92	0.0014
H_2O	3.06	18.01	0.1699
Total	100.59		

Calculation of F
$F = 0.0561 - 0.0118 - 0.0194 = 0.0249$ $F\% = 0.0249/0.1206 \times 100 = 20.6\%$
Calculation of A
$A = 0.1553 - 3 \times 0.0297 - 0.0392 = 0.0270$ $A\% = 0.0354/0.1290 \times 100 = 22.4\%$
$A + F + M = 0.0249 + 0.0354 + 0.0687 = 0.1290$

*Column 1: weight percentages of oxides in garnet–mica schist from Alasehir, Turkey. Analy-
      sed by R. Akkok.
Column 2: Molecular weights of oxides, from Appendix.
Column 3: Molecular proportions, see text.
†Mineral assemblage of this rock: garnet + albite + biotite + muscovite + quartz + chlorite +
apatite + zircon + sphene

Sometimes [MnO] is added to the [FeO] total in the calculation of F. This is because Mn^{+2} substitutes for Fe^{+2} in many silicate minerals. This may occasionally lead to problems in representing the mineral assemblages if Mn-rich garnet is present and has therefore not been recommended here. The proportion of [MnO] in most pelitic rocks is small, so the differences in the position plotted on the diagram between including it and leaving it out are correspondingly small.

Mineral compositions on ACF and AFM diagrams

Any one mineral analysis will plot as a point on a composition–assemblage diagram, like a rock analysis. The chemical formula given for a mineral represents an ideal composition, and real analyses differ because one element may substitute for another in the crystal structure. Thus almandine garnet has the ideal formula $Fe^{+2}_3Al_2Si_3O_{12}$, and would thus plot on the right-hand edge of an ACF triangle. Real garnets, however, may have Ca substituting for Fe, and may thus plot farther to the left. Figure 2.16a shows an ACF

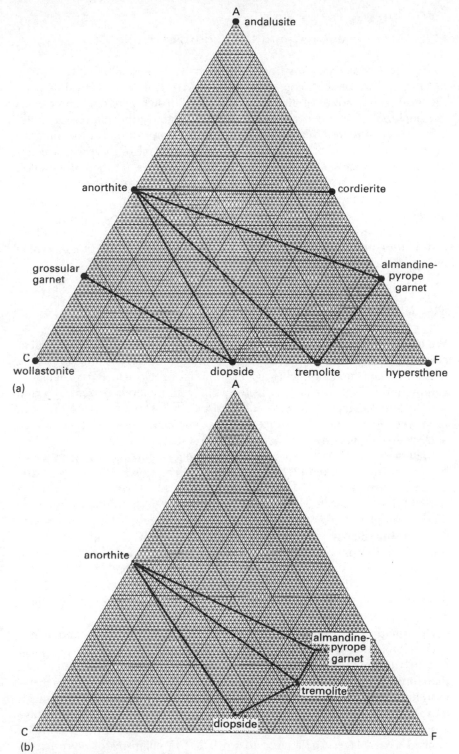

Figure 2.16 (a) ACF diagram for skarns from Neyriz, Iran, plotting 'ideal' compositions for minerals. (b) The same diagram, plotting actual mineral compositions as dots. Notice that there are fewer minerals on this diagram, because the only minerals plotted are those which have actually been analysed. From data supplied by R. Hall.

triangle plot of analyses of garnet, clinopyroxene and amphibole from some metamorphosed basic igneous rocks from Neyriz, Iran. The heavy dots show the ideal compositions of almandine garnet, diopside pyroxene and tremolite amphibole. The smaller dots represent the actual compositions of analysed minerals. Figure 2.16a shows the idealized compositions of the minerals joined by tie-lines, representing the mineral assemblage of this rock, which is garnet + pyroxene + amphibole. Figure 2.16b shows the actual tie-lines, linking measured rock compositions. The patterns and positions of tie-lines alter with changing metamorphic conditions, and therefore such diagrams are widely used to describe changes in mineral assemblages.

The ACF and AFM diagrams presented here are only two of many possibilities, and other types of diagrams will be used in this book, and may be found in other publications.

Metamorphic textures

We have seen that the principles of metamorphism suggest that the number of types of mineral present in a metamorphic rock is likely to be small, about five or six major minerals being the maximum for rocks which have achieved equilibrium. While it is important to identify all minerals present in a thin section of a metamorphic rock, and to decide whether or not they crystallized together, metamorphic textures (or **microstructures**) should not be neglected when making systematic petrographic descriptions.

Metamorphic textures reflect (1) the original textures and mineral composition of the rock and (2) the growth and break down of crystals during metamorphism. Figure 7.11 shows a metamorphosed turbidite, in which the size and shape of quartz and feldspar crystals are inherited from the unmetamorphosed clastic sediments. The mica and chlorite crystals, by contrast, grew during metamorphism accompanied by deformation, and mainly reflect metamorphic processes. This simple difference between types of mineral grain in a single rock sample is unusual, the common thing being for grains to show the effects of both the primary igneous or sedimentary processes which formed the rock, and also the metamorphic processes which have subsequently occurred.

Deformation processes have particularly characteristic effects in dynamic and regional metamorphic rocks. The mechanisms of deformation are sensitive to the conditions of metamorphism, particularly temperature, and therefore an estimate of the P–T–t path followed by a rock during metamorphism is important for microstructural study. In many suites of metamorphic rocks, a two-stage approach is the most productive. The first stage is to establish the sequence of metamorphic conditions from a study of mineral assemblages, the second to work out mechanisms of deformation from an analysis of metamorphic textures. The textures may enable us to fix

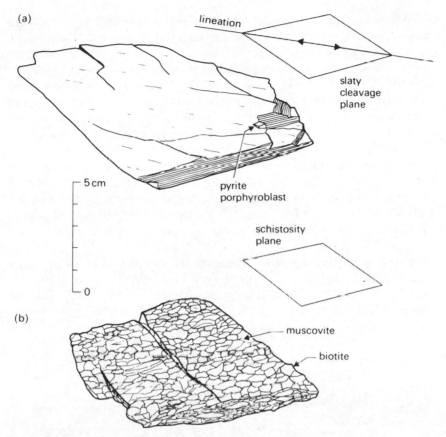

(a)

lineation

slaty
cleavage
plane

5 cm

pyrite
porphyroblast

schistosity
plane

0

(b)

muscovite

biotite

Figure 2.17 Hand specimens of (a) slate and (b) schist, to illustrate the more regular cleavage direction in slate. Individual muscovite and biotite crystals are distinguished in the schist. Slate from Ballachulish, Argyll; schist from Lock Stack, Sutherland, Scotland.

the times of significant events on the P–T–t track, relative to times of deformation events such as folding or overthrusting. In some cases, this leads to the formulation of an integrated tectonic model, in which the changes in temperature, pressure and type of deformation through time can be reconstructed.

Textures in which mineral grains show a tendency to lie in certain directions (i.e. preferred orientations) are characteristic of many regional metamorphic rocks. Textures formed by the combination of tectonic deformation with metamorphism are often referred to as **metamorphic fabrics**, and the preferred orientations of mineral grains as **directional fabrics**. There are two main geometrical classes of directional fabrics – planar fabrics or foliations and linear fabrics or lineations. The two may be combined.

Foliation fabrics (Fig. 2.17) are those in which the fabric elements, usually mineral grains, have a planar arrangement. An alternative name is

lepidoblastic texture. One familar foliation fabric is **slaty cleavage**. Slates are fine-grained rocks of pelitic composition, largely composed of phyllosilicate minerals such as white micas and chlorite. The platey crystals of these minerals are arranged so that they are approximately coplanar, or loosely speaking, parallel (Fig. 2.18). Because phyllosilicate minerals have a good cleavage coplanar with the platey crystals, the rock splits very easily into thin sheets in the direction of the crystals. This property is known as **rock cleavage** to distinguish it from the mineral cleavage of individual crystals. If there is no danger of confusion, it is just known as cleavage.

Coarser-grained metamorphic rocks of pelitic composition also have their individual crystals arranged in a coplanar fabric. Although rock cleavage is still present, the resulting fracture surfaces are less perfectly flat and the sheets which can be split from the rock are thicker, because the crystals are larger, and less perfectly coplanar. The cleavage in these rocks is called schistosity (Fig. 2.19).

The origin of cleavage and schistosity is complex. The compaction of mud rocks, which are made up of very small crystals of phyllosilicate minerals, changes the original random arrangement of the crystals into an arrangement where they lie parallel to one another, and parallel to the bedding direction. Tectonic stress may re-orient the crystals, and they also

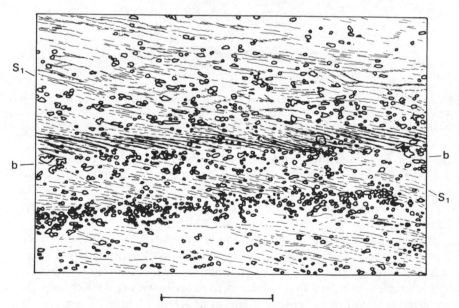

Figure 2.18 Thin section of slate, Manod Quarry, Blaenau Ffestiniog, Gwynedd, Wales. The rock is too fine grained for individual crystals of muscovite and chlorite to be distinguished at this scale, so their general direction is indicated by fine lines, which represent concentrations of crystals of < 0.01 mm size. s_1–s_2 cleavage direction, b–b bedding direction. Compare with Figure 7.11. Scale bar 1 mm.

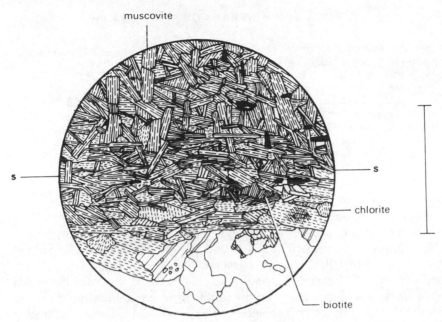

Figure 2.19 Thin section of schist, from below the garnet isograd, Sulitjelma, Norway. Scale bar 1 mm.

Figure 2.20 Gneiss showing almost horizontal metre scale banding. Lake Placid Ski Center, Adirondack Mountains, New York, USA

Figure 2.21 Granitic augen gneiss, Ossola Valley, Italian Alps. The crystals with oval cross sections, elongated in the lineation direction, are of K-feldspar, and were originally phenocrysts in a granite. The groundmass between them has been converted into a granitic mylonite. Compare Figure 4.7. Height of sawn block 1 cm.

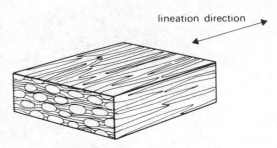

lineation direction

grow larger due to metamorphic recrystallization (pp. 33–36).

Another common type of foliation fabric is a banding of the rock into layers rich in ferromagnesian minerals, such as biotite and hornblende, alternating with layers rich in quartz and feldspars (Fig. 2.5). This type of banding is seen in the large class of metamorphic rocks known as gneisses, and will be called gneissose banding in this book. It is sometimes called gneissosity. The bands may vary in width, from less than a millimetre to several metres (Fig. 2.20). The gneissose banding is only detectable in thin section in the more finely banded varieties. In most gneisses, the micas of the ferromagnesian layers are also parallel to one another, i.e. they have a schistosity.

Lineation fabrics are less familiar than foliation fabrics. In rocks with prismatic or acicular mineral grains these may have a parallel arrangement (the word 'parallel' is not being used loosely this time). Figure 2.21 is a sketch of a hand specimen of gneiss from the Ossola Valley, in the Italian Alps, which shows a lineation fabric defined by elongate crystals of feldspar.

The systematic study of the relationships between geological structures and metamorphic fabrics leads to an understanding of the time relationships between metamorphism and deformation. This can be illustrated by considering the relationship between slaty cleavage and folds in low-grade regional metamorphic rocks. In such rocks the cleavage direction is commonly coplanar with the axial surfaces of the folds (Fig. 2.22). Interpretations of the mechanisms of formation of folds suggest that the cleavage was developed and the folding occurred over the same period of time. If the time of folding can be determined, or more usually fixed within a certain timespan, then the time when the cleavage evolved is also known. There is a problem to this approach, which is that once the minerals of a rock have taken on a preferred orientation, they may retain it, while their compositions are modified by later metamorphic events. So if the minerals defining the slaty cleavage or schistosity attained equilibrium at a certain temperature and pressure, these are not necessarily the conditions under which the folding occurred. They could have been 'set' in the mineral assemblage later.

Metamorphic facies

The idea of metamorphic facies builds upon the recognition of mineral assemblages coexisting in equilibrium, by emphasizing that any mineral assemblage will coexist over a particular range of metamorphic conditions, temperature and pressure being the two most important. This idea was first propounded by the great Finnish petrologist, P. Eskola in 1915, before there were any significant experimental or thermodynamic data on metamorphic minerals. In 1920, Eskola listed a number of diagnostic mineral assemblages in rocks of basic igneous composition, defining broad ranges of temperature and pressure, although at that time it was not possible to give numerical values to plot onto P–T diagrams. Later workers have considered wider ranges of rock composition, and more possible variables during metamorphism (such as the chemical activities of H_2O and CO_2), and thus tended to propose a larger number of metamorphic facies than in Eskola's original scheme, in which there were only nine. In all facies classification schemes, individual facies are defined by specifying the diagnostic mineral assemblages for rocks of common bulk compositions, such as basic igneous rocks and pelitic rocks. Individual metamorphic facies are given names after

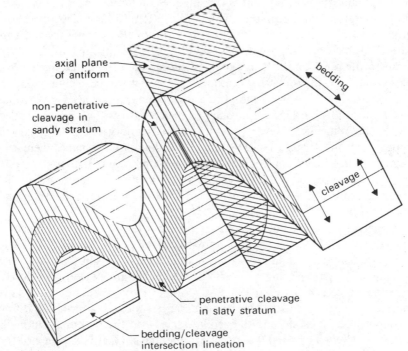

Figure 2.22 Illustration of the geometrical relationships between axial-plane cleavage and bedding.

common metamorphic rock type stable under the appropriate conditions (e.g. greenschist, eclogite), and in some schemes subdivisions of metamorphic facies may be called after diagnostic mineral assemblages.

An individual metamorphic facies may be defined in one of two ways. Diagnostic mineral assemblages may be defined for particular rock compositions (e.g. assemblages with glaucophane + lawsonite in basic igneous rocks for the blueschist facies). Eskola originally formulated nine metamorphic facies on this basis – pyroxene hornfels, sanidinite, greenschist, epidote–amphibolite, amphibolite, granulite, glaucophane–schist and eclogite. Alternatively metamorphic reactions may be taken to define the boundaries of the metamorphic facies (e.g. muscovite + quartz = potash feldspar + H_2O for the boundary between the amphibolite facies and the pyroxene hornfels facies). This approach has gained in popularity as more becomes known about the P–T conditions for individual metamorphic reactions. If we could identify all the metamorphic reactions which could occur in a rock of a particular composition, and knew all the thermodynamic properties of the participating minerals and fluid phases, there would not be any difference between the two approaches, but we are not in such a favourable position.

Another feature of the metamorphic facies classification of rocks is that it treats metamorphism as a process which goes on under one set of conditions of temperature and pressure. These are often loosely described as the 'pressure and temperature of metamorphism' or 'P–T conditions'. In more careful discussion, they will be taken as the conditions at the highest temperature point on the P–T–t path, often called the **metamorphic peak**. For many metamorphic rocks, for example the contact metamorphic rocks which Eskola considered in his original account, the assumptions are reasonable, and a metamorphic facies approach to classification is a very fruitful one. Under the influence of several textbooks on metamorphic petrology, which appeared in the 1960s and 1970s, metamorphic facies schemes were widely applied to link metamorphic conditions with plate tectonic settings during the pioneering period of the development of the plate tectonic theory. So an understanding of the schemes is needed to read important scientific papers from that period.

Modern studies of metamorphic rocks have been strongly influenced by the works of F. J. Turner, beginning with a clear exposition of the principles of metamorphic facies and its experimental foundations in Fyfe et al. (1958) and culminating in Turner's own book 'Metamorphic petrology' (1981). He explained the link between the P–T conditions of metamorphism, metamorphic reactions and mineral assemblages, and recognised 11 different metamorphic facies, adding the zeolite facies and the prehnite–pumpellyite facies to Eskola's original 9. Another influential text, advocating a different facies scheme, was 'Petrogenesis of the metamorphic rocks' by Winkler, first published in 1967 with further editions until Winkler (1986). He originally

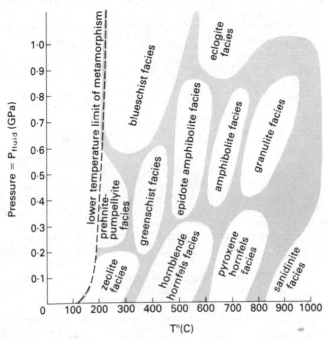

Figure 2.23 Approximate pressure and temperature ranges of different metamorphic facies, based on Miyashiro 1973. Shaded areas represent P–T conditions which are transitional between different facies.

used facies classifications with a large number of divisions into facies and subfacies, but in his later editions adopted a broad subdivision of P–T conditions into different 'grades', mainly characterized by different temperatures of metamorphism. The fullest modern account of the facies approach is in the Soviet compilation 'The facies of metamorphism' edited by Sobolev (1972), which uses a broad division into facies similar to those of Eskola and Turner, with divisions into subfacies, often illustrating the P–T conditions by reference to rocks from the USSR.

The Japanese petrologist Miyashiro, who pioneered the application of the study of metamorphic rocks to tectonic problems, emphasized facies *series* rather than individual facies in his book 'Metamorphism and metamorphic belts' (1973). He demonstrated that sequences of metamorphic facies are often found in metamorphic belts, for example prehnite–pumpellyite to blueschist to greenschist, and greenschist to amphibolite in the paired metamorphic belts of Japan. He related these facies sequences to the different geothermal gradients occurring near subduction zones.

This book will refer to a broad facies scheme, based on Miyashiro (1973) and Turner (1981), but is not constructed around the scheme. The fields of temperature and pressure of individual facies are shown in Fig. 2.23, and

Table 2.4 Names of metamorphic facies and typical mineral assemblages of basic rocks and pelitic rocks in each.

Facies name	Typical assemblage in basic igneous rocks	Typical assemblage in pelitic rocks
Facies of regional metamorphism		
zeolite*	smectite + zeolites (with relict igneous plagioclase and pyroxene)	illite + chlorite + quartz
prehnite–pumpellyite*	prehnite + pumpellyite (with relict igneous plagioclase and pyroxene)	illite + stilpnomelane + chlorite + quartz
greenschist	chlorite + actinolite + albite + epidote + quartz	muscovite + chlorite + chloritoid + quartz
epidote–amphibolite	hornblende + albite + epidote + garnet + quartz	muscovite + biotite + garnet + quartz
amphibolite	hornblende + andesine + garnet + quartz	biotite + muscovite + garnet + sillimanite + quartz
granulite	labradorite + clinopyroxene + orthopyroxene + quartz	biotite + garnet + cordierite + sillimanite + quartz
blueschist	glaucophane + lawsonite + albite (or jadeite) + quartz	muscovite + chlorite + spessartine garnet + quartz
eclogite	pyrope garnet + omphacite	muscovite + kyanite + biotite + garnet + quartz
Facies of contact metamorphism		
hornblende hornfels	hornblende + andesine + quartz	biotite + muscovite + cordierite + andalusite + quartz
pyroxene hornfels	clinopyroxene + labradorite + quartz	biotite + muscovite + cordierite + andalusite + quartz
sanidinite	clinopyroxene + labradorite + quartz	sanidine + mullite + hypersthene + cordierite + quartz

*Basic igneous rocks of the zeolite and prehnite–pumpellyite facies seldom have an equilibrium mineral assemblage. Common relict igneous minerals are therefore listed in brackets.

Table 2.4 lists common (but not necessarily diagnostic) mineral assemblages for metamorphosed basic igneous rocks and for pelitic rocks in each facies. The P–T fields of the metamorphic facies are indicated roughly on Fig. 2.23, but exact limiting boundaries are not given. This is because there is little agreement, either on the specific metamorphic reactions which should be taken as the boundaries of facies, or on the P–T curves for the reactions concerned. Table 2.4 is not intended to be a comprehensive guide to metamorphic facies and mineral assemblages, because although facies is a useful guide to broad P–T conditions of metamorphism, it has become clear in recent years that it will not provide a set of 'pigeon-holes' which will describe the P–T conditions of all metamorphic rocks. Its use as an intermediate step between the lists of minerals in a particular rock, and the P–T conditions of metamorphism is being supplanted by more direct methods of determining P and T. This book therefore does not recommend students to regard the assignment of rocks to a particular metamorphic facies as a major aim of study, although you should recognise that many experienced petrologists do favour this approach.

Thermotectonic modelling

In Chapter 1, the relationship between geothermal gradients and field metamorphic gradients was briefly explored. This section explores the scientific principles behind this linkage, in order to provide a framework for the discussion of areas of metamorphic rocks later in the book (England & Thompson 1984).

The interior of the Earth is hotter than the surface. One expression of this fact is the increase in temperature observed in deep mines and boreholes (Bott 1982). Within the Earth's rigid outer layer, the **lithosphere**, heat escapes from the interior by conduction upwards through the rocks of the crust and upper mantle. The temperature gradient is related to the rate of flow of heat from deeper levels by the equation for heat conduction, which is

$$dQ/dt = k.dT/dz \qquad (1)$$

dQ/dt is the rate of flow of heat, per unit area, measured in microwatts per square metre ($mW\ m^{-2}$), and dt/dz the rate of increase of temperature with depth, measured in °C per metre or per kilometre. k is a constant known as the **thermal conductivity** characteristic of the rocks through which the heat flows. It can be measured in laboratory experiments, and studies on individual boreholes show an excellent agreement between the actual thermal gradients measured in the boreholes, and those calculated from equation (1) applying values for k measured on the samples taken from the boreholes (Bott 1982). There are qualifications to this observation. For example if the rocks contain pore fluid such as water, which can flow through them (i.e. they are permeable), heat may be transferred by flow of fluid, and equation (1) is no longer applicable. However, at the depths and temperatures of interest for metamorphism, equation (1) has wide application.

The heat originates in the interior of the Earth in three forms:

(1)   Energy released by nuclear reactions, i.e. the decay of unstable isotopes
      of chemical elements occurring in the interior of the Earth.
(2)   Residual heat surviving from a period shortly after the accretion of the
      Earth from the planetary nebula, when it was much hotter than it is
      now.
(3)   Local heating of the deeper layers in the Earth, caused by a variety of
      volcanic and plate tectonic processes.

   The relationship between heat flow from the interior and thermal gradients in the lithosphere is fundamentally different on ocean floors and in
stable continental areas. Oceanic heat flow will be discussed in Chapter 5,
and continental heat flow will be discussed here in order to illustrate general
principles.
   Continental areas have been stable over very long periods of geological
time, > 1000 Ma. Heat-flow measurements in such areas show that the rate
of heat flow dQ/dt tends to a constant value of about 50 mW m^{-2} in the
oldest continental areas. In such areas, the effects of local changes in heat
flow, listed as (3) above, have died away. They have achieved a steady state
of heat flow. At deep levels in the crust, such stable continental areas are
usually composed of gneisses of Archaean or Proterozoic age, and the
average thermal conductivity of these is quite uniform, so a uniform geothermal gradient is also achieved. This is called the steady-state continental
geothermal gradient. Should the thermal gradient in an area of continental
crust be disturbed for example by a period of volcanic eruption and
intrusion, the steady-state geothermal gradient will change for a while, but
when the volcanic activity has ceased, gradually as time passes it will
tend to return to the steady-state value.
   The steady-state geothermal gradient is not uniform with depth. The
granitic rocks present in continental gneisses contain a relatively high proportion of unstable isotopes of potassium, thorium and uranium which
supply heat within the crust, making dQ/dt larger at shallow levels than at
deeper levels. As a result, the geothermal curve is convex upwards on a
temperature/depth plot (Fig. 1.5). The fact that temperature/pressure plots
for metamorphic gradients in continental areas are also convex upwards
(Fig. 1.6) suggests that a comparison of thermal gradients in the lithosphere
with metamorphic gradients is reasonable.

A simple example

Let us consider what will happen if the heat flow dQ/dt is changed. We can
take the example of an old area of continental crust, where the geothermal
gradient has attained the steady-state value, and suppose that igneous
intrusions into the lower crust raise dQ/dt from 50 mW m^{-2} to
100 mW m^{-2}. We shall discuss a simplified case, ignoring the production of

heat within the crust and variations of thermal conductivity k with depth or temperature. This makes our geothermal curves become straight lines. We can use equation (1) to calculate the higher geothermal gradient if the continental crustal area achieves a steady-state geothermal gradient at the higher heat-flow value. When the igneous activity has ceased, the heat flow will return to 50 mW m^{-2}, and the geothermal gradient will relax to its previous value. The thermal gradients are shown in Fig. 2.24.

Now let us consider two rocks in the portion of the continental crust which we have been discussing, one at a shallow level, one deeper down. Fig. 2.25 shows the change of temperature with time in each rock, assuming that there has been enough time for thermal gradients to come close to their steady-state values. It can be seen that the shallow rock has undergone a small increase in temperature, sustained over quite a long period. The deeper rock has undergone a considerable increase in temperature, sustained over a long period of time. Both are very likely to display metamorphic effects as a result. An increase in heat flow due to intrusion of igneous material deep in the crust is a very likely cause of metamorphism, and in Chapter 6 we shall discuss an example where precisely this has occurred.

This preliminary discussion has shown that thermotectonic modelling can predict the pressure–temperature history of rocks within the Earth's lithosphere, using simple physical theory and assumptions about the order in which geological events occur. Later chapters will introduce thermotectonic models for different types of plate boundaries, such as oceanic ridges and subduction zones. All the later discussions share a number of assumptions with this first model:

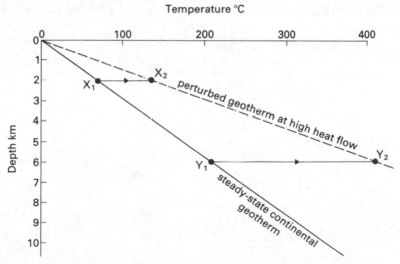

Figure 2.24 Schematic geothermal gradients from example discussed in the text. X_1–X_2 change in temperature of rock 2 km below surface, Y_1–Y_2 change of temperature in rock 6 km below surface.

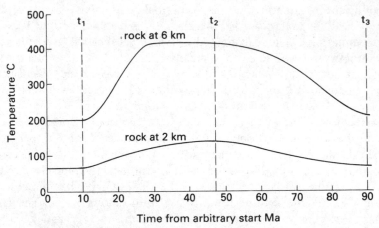

Figure 2.25 Temperature–time curves from example discussed in the text. Upper curve, rock at 6 km depth, lower curve, rock at 2 km depth.

(1) In parts of the lithosphere which have been stable for more than 100 Ma, which are all continental shield areas or platform areas, the rate of flow of heat from the Earth's interior has settled to a constant value. This produces a similar geothermal curve beneath the surface in all such areas, which is known as the **steady-state continental geotherm**.

(2) Tectonic and volcanic events disturb this heat-flow pattern, giving rise to **perturbed geotherms**, which may show higher or lower geothermal gradients than the steady-state continental geotherm.

(3) When tectonic and/or volcanic activity ceases, the perturbed geotherms relax towards the steady-state geotherm. This process takes a relatively long time, in the range of 10–50 Ma.

3 Metamorphism associated with igneous intrusions

The processes involved in metamorphism are probably most easily understood in cases where surrounding country rocks have been heated by an igneous intrusion. It is increase in temperature which causes the most striking effects on the mineral assemblages in a rock, and in these cases of **contact metamorphism**, the source of the heat is obvious. It is the heat originally contained in the molten liquid magma of the intrusion. Perhaps the simplest case of all is that of a sheet of basic magma intruded into a planar fracture in the Earth's crust, which has cooled to form a dyke of dolerite.

There is a zone in the country rock adjacent to each margin where the rock

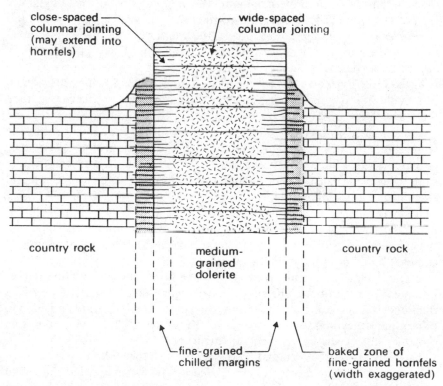

Figure 3.1 Cross section through a dolerite dyke 1 m wide, showing contact metamorphism of country rock.

has been changed by metamorphism. If the country rock is sedimentary (for example a shale), for a dyke about 1 m thick there might typically be a zone 2 mm wide of fine-grained rock, tougher and more resistant to weathering than the surrounding unaltered rock. When this rock is broken, the exposed surfaces resemble the broken surfaces in pottery. Since the basic magma was very much hotter than the country rock at the time of the intrusion, perhaps 1000°C compared with 30°C, the alteration of the country rock has obviously been caused by the strong heating occurring during the intrusion, and immediately afterwards while the dyke cooled. The narrow zone of porcellaneous rock is called the 'baked margin' of the country rock, corresponding to the chilled margin of the dyke itself (Fig. 3.1). In the case of a vertical dyke, intruded into impermeable country rocks, the heat for metamorphism was mainly transferred across the contact by conduction through the country rocks. Soon after the onset of intrusion, the heat would also have to be transferred by conduction through the solidified magma in the chilled margin.

Contrary to what we might think, the temperature of the country rock next to the contact usually does not rise to the temperature of the intruding magma. This is because the magma at the very edge of the dyke instantly cools and solidifies as it comes into contact with the cold country rock, and thereafter heat from the hotter molten magma inside the dyke must travel through the solidified igneous rock by conduction. This conclusion only applies if the dyke filled quickly with hot magma, which then cooled in place. If instead, the dyke acted as a 'feeder' to a higher lava flow or intrusion so that fresh, hot magma kept passing the chilled margins up the centre, then the temperature of the country rocks would rise to the temperature of the magma, exactly as the temperature of a central heating radiator rises to the temperature of the hot water passing through it. There is geological evidence to support the difference between these two cases, with examples of intrusions which acted as feeders having unusually high-temperature mineral assemblages at their contacts.

Away from the immediate contact at a time shortly after intrusion, the temperature will fall off rapidly with increasing distance, but as time goes on, the temperature gradients will tend to flatten out. Fig. 3.2 shows a series of curves, plotted at different times t_1, t_2, t_3. . . . after intrusion (Jaeger 1957). The maximum temperature is always at the contact itself, but the curves become flatter with time. The curves in the figure have been calculated on a theoretical basis, using the equation for conduction of heat, and among other things they indicate that the larger an intrusion is, the wider the zone on either side which is heated by the intrusion.

This conclusion can be tested by observation of the width of the zones of metamorphism around intrusions. A dolerite dyke 1 metre thick intruded into cold country rocks develops a baked margin 1–2 mm thick. The Whin Sill of northern England, also of dolerite, is about 73 m thick, and has a baked

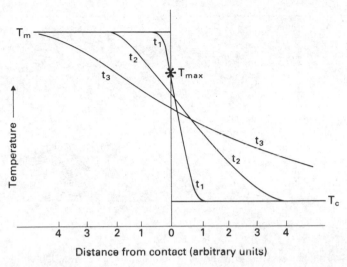

Figure 3.2 Variation of temperature with time across the contact of an igneous intrusion, such as the dyke in Figure 3.1, in which there is a single, quick intrusion of magma. T_m (horizontal line) – temperature of magma at time of intrusion (t_0), T_c (horizontal line) – temperature of country rock at t_0, T_{max} (asterisk) – maximum temperature of country rocks at the contact, a very small time after intrusion. t_1, t_2, t_3 – temperature–distance curves for later times.

margin up to 40 m wide, above and below. The massive and fine-grained country rock in this baked margin is harder than the unmetamorphosed sediments. It tends to display a conchoidal fracture and to shatter into numerous sharp-edged fragments. Under the microscope, it is seen to be composed of small crystals of calcite, with planar grain boundaries and 120° triple junctions, i.e. a fine-grained granoblastic texture. This type of tough, massive rock found adjacent to the contacts of igneous intrusions is called hornfels.

In larger intrusions still, the baked margins become even wider. The Sulitjelma gabbro, Norway, is an irregular, sheet-like intrusion, which may have been formed by numerous pulses of intrusion of basic magma (Mason 1971, Boyle 1989). It has a baked margin of massive hornfels 30–100 m wide, where it is intruded into regional metamorphic rocks (Chapter 7). The hornfelses are relatively easily distinguished from the regional metamorphic rocks in the field. They are fine-grained and massive, whereas the regional metamorphic rocks are coarser-grained and often split easily into parallel sheets because of the preferred orientation of the phyllosilicates. The horn-felses have no preferred orientation fabric, while the regional metamorphic rocks display schistosity. The hornfelses are more resistant to erosion than either the gabbro or the country rocks, and therefore in many places a topographic ridge marks the position of the contact. This is quite a frequent feature of large intrusions surrounded by hornfelses.

Under a hand lens, and more distinctly with the petrological microscope,

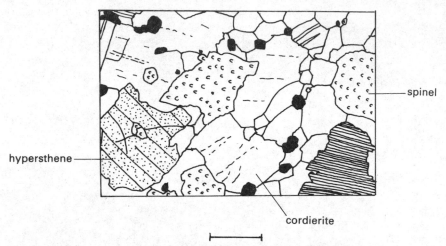

spinel

hypersthene

cordierite

Figure 3.3 Hornfels showing equilibrium granoblastic texture with 120° triple junctions. Contact aureole of Sulitjelma gabbro, Norway. Scale bar 0.1 mm.

the Sulitjelma hornfelses can be seen to have a rather uniform grain size, and the grains themselves to approach equant shapes (except for the biotite mica flakes) (Fig. 3.3). They are coarser grained than the hornfelses at the contacts of the Whin Sill, and the granoblastic texture gives the hornfels an even-grained, granular appearance when it is studied under a hand lens.

As the zone of baking around intrusions becomes larger, it is possible to recognize a gradient of increasing metamorphic change as the contact is approached. Contact metamorphic rocks are found surrounding granite and granodiorite intrusions, as well as dolerite and gabbro intrusions. While basic magma predominates in small intrusions, silica-rich magma predominates in larger ones, and therefore examples of contact metamorphism more often occur surrounding large intrusions of intermediate or silica-rich magma.

The baked rocks in wider zones of contact metamorphism often show an increase in grain size as the contact is approached, although there is a lot of variation in the grain sizes of hornfelses both in different places around any particular intrusion and between different intrusions of the same size. Large zones of contact metamorphic rocks surrounding intrusions are called **contact aureoles**. The name is applied to any zone of contact metamorphic rocks that is wide enough to be represented on a geological map. Large intrusions of granitoid composition may have contact aureoles more than a kilometre wide (Fig. 3.4).

The sequence from unmetamorphosed country rock to the most metamorphosed hornfels immediately against the contact of the intrusion is called a **progressive metamorphic sequence**. Particularly impressive sequences of textural and mineralogical changes are seen in progressive contact metamorphic sequences in which the country rocks are **pelitic rocks** (Chapter 2).

A good example of a progressive contact metamorphic sequence in pelitic rocks is seen in the contact aureole of the Markfield Diorite in Cliffe Hill Quarry, Leicestershire, England (Evans *et al.* 1968). The country rock is a late Proterozoic slate of the Woodhouse Eaves Formation. Diorite and hornfels are quarried for road-surfacing aggregate, both serving equally well for this purpose. Consequently, the contact between diorite and hornfels is exposed in the working faces and floor of the quarry, allowing the progressive contact metamorphic sequence to be studied in unweathered rocks over a distance of 40–50 m. The relationships of the different rock types are shown schematically in Figure 3.5.

The first recognizable change due to contact metamorphism is that the

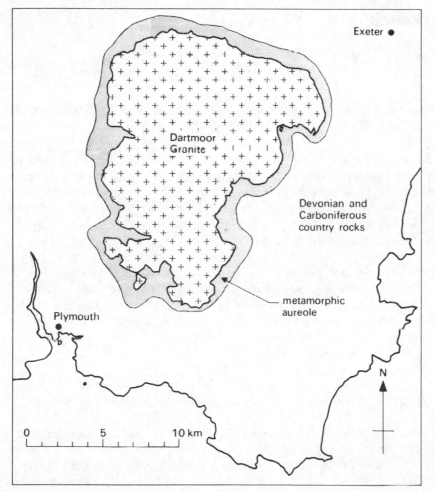

Figure 3.4 The contact aureole surrounding the Dartmoor Granite, Devon, England. Based upon British Geological Survey map sheets, with permission.

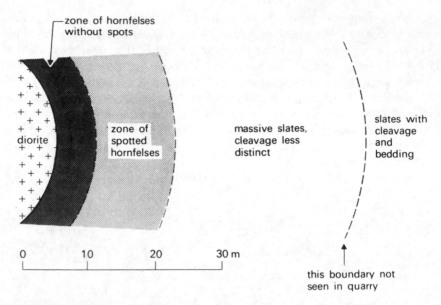

Figure 3.5 Schematic plan of the contact metamorphic zones seen in Cliffe Hill Quarry, Markfield, Leicestershire, England.

slates of the Woodhouse Eaves Formation lose their tendency to split (i.e. their slaty cleavage), and become massive. This change is not seen inside the quarry itself where the rocks continue to become more massive as the contact is approached. Between about 5 m and 25 m from the contact, dark spots appear along some of the original bedding surfaces of the country rock, and in the last 5 m up to the contact, the rock is very hard and massive, to become a true hornfels. It shows banding from light grey to dark grey, but the spots are not present. The banding can be identified as preserved sedimentary bedding, by original sedimentary structures such as channels and ripple marks seen on the bedding surfaces. There are also rain pits and impressions of the late Proterozoic fossil *Charniodiscus* preserved on some bedding planes in the massive hornfels zone (Evans *et al.* 1968). The sequence may be summarized:

slate →   massive slate   →   spotted hornfels → diorite
(country rock) → (progressive contact metamorphic sequence) → (intrusion)

The unusual degree of preservation of primary sedimentary features in hornfelses close to the contact has been recorded for other contact aureoles (Craig 1983). The development of spots is a characteristic feature of contact metamorphic rocks of pelitic composition, which are called **spotted slates**.

The progressive metamorphic sequence seen in the contact aureole of the Markfield Dolerite may thus be described as one in which metamorphism was most intense adjacent to the contact, and died away at greater distance from the contact. This may be expressed more simply by saying that the metamorphic grade of the rocks is high at the contact, and becomes steadily lower towards the outer limit of the contact aureole. In the field, the term 'metamorphic grade' is used in a loose sense for describing the intensity of metamorphism, 'high grade' describing the more intensely metamorphosed members of a progressive metamorphic sequence, 'low grade' the less intensely metamorphosed members. Grade may be used in this way to describe the degree of metamorphism in contact, dynamic or regional metamorphic rocks.

In our present contact metamorphic example, the rocks next to the contact of the diorite are at high grade, and have also been to the highest temperatures during metamorphism (Fig. 3.2). Some accounts of metamorphism equate metamorphic grade with the temperature of metamorphism (Winkler 1986), but closer study of Figure 3.2 suggests that the length of time available for metamorphism is also important in this example. Figure 3.6 plots the same results as those in Figure 3.2 on a time–temperature graph, for only two rocks, one adjacent to the contact, and one at some distance away. It shows that the rock near the contact remained close to its maximum temperature for a relatively long time. The one away from the contact reached its maximum temperture appreciably later, and remained at a high temperature for a shorter period. Some metamorphic effects, such as the tendency for the grain size to become larger, are more time-dependent than temperature-

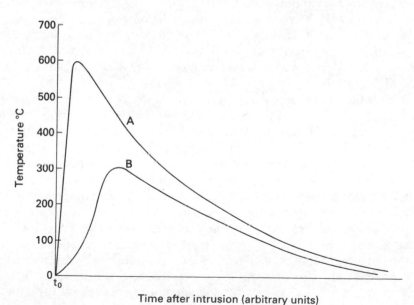

Figure 3.6 Temperature–time curves for a rock close to an intrusion (Curve A), and further away from the intrusion (Curve B).

dependent, and thus to equate metamorphic grade (used as a descriptive field term) with metamorphic temperature is a misleading oversimplification.

In many contact metamorphic sequences it is possible to define stages between low and high grade by observing the entry of new minerals, as will be described later in this chapter. However, contact metamorphic aureoles are more usually subdivided into zones by textural changes in the rock.

In the inner parts of some contact aureoles, the country rocks have not only been recrystallized during metamorphism, they have also undergone a change in chemical composition. In all aureoles, the proportion of volatile components of the rock such as CO_2 and H_2O tends to decrease with increasing metamorphic grade. But in some aureoles the proportions of less volatile components such as potassium, calcium and boron also change. Metamorphism involving changes in rock composition of this sort is called **metasomatism**, and in cases where it occurs at the contacts of igneous intrusions it is called 'contact metasomatism'.

The Skiddaw contact aureole, Cumbria, England

In the northern part of the English Lake District, in the county of Cumbria, several small areas of outcrop of biotite granite or granodiorite are seen in contact with the Ordovician Skiddaw Slate Formation (Moseley 1978). They are surrounded by a wide contact aureole, which is the type example which has been presented to British students for many years, having been described by Rastall in 1910. Geological mapping of the outcrops of granitoids and their contact aureole (Eastwood *et al.* 1968) indicate that the granitoids are joined together at a shallow depth below the land surface, and represent the irregular domed roof of a medium-sized batholith (Fig. 3.7) of Devonian age (390 Ma, Soper 1987).

Outside the contact aureole, the slates have a cleavage which is parallel to the axial surfaces of tight folds. Cleavage and folds are overprinted by the contact metamorphic aureole, indicating that they belong to a pre-intrusive episode of rock deformation and accompanying regional metamorphism. Therefore the Skiddaw Aureole shows contact metamorphism superimposed on low-grade regional metamorphism. Microscopic study of the Skiddaw slates outside the contact aureole reveals the mineral assemblages associated with the low-grade regional metamorphism. The phyllosilicate minerals are muscovite and chlorite, with a strong preferred orientation parallel to the slaty cleavage direction. Grains of quartz retaining their original clastic shapes, small prisms of chloritoid and the opaque minerals pyrite, magnetite and graphite are also present. These are characteristic minerals of aluminium-rich black shales which have undergone regional metamorphism.

The progressive contact metamorphic sequence in the pelitic rocks of the Skiddaw Aureole is divided into three zones. The sequence, in order of

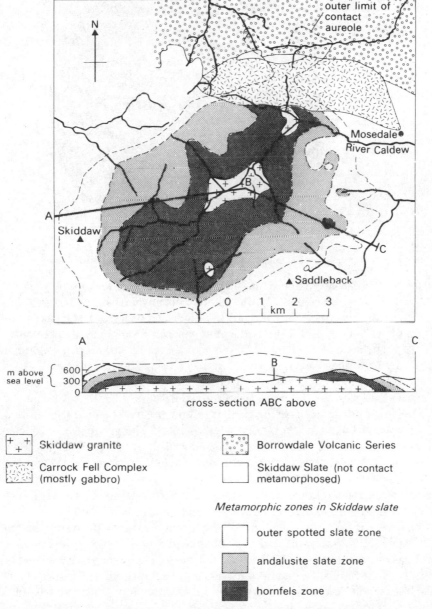

Figure 3.7 Geological map of the Skiddaw Granite and its contact aureole, Cumbria, England, from Eastwood *et al.* (1968).

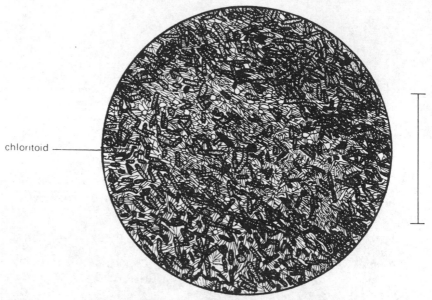

chloritoid

Figure 3.8 Spot in spotted slate, outer zone of the Skiddaw aureole. Scale bar 1 mm.

increasing metamorphic grade is: unaltered Skiddaw slates (country rock), (1) outer spotted slate zone, (2) intermediate porphyroblastic slate zone, and (3) inner hornfels zone. The zones are distinguished on a textural basis rather than on variation in the metamorphic mineral assemblages. The outer spotted slate zone (1) is characterized by a coarsening of the grain size and a tendency for the cleavage surfaces to become more uneven. Spots 0.15–2.0 mm across, composed of white mica and chloritoid, appear on the cleavage surfaces. In the porphyroblastic slate zone (2), the spots can be identified as porphyroblasts of cordierite and prisms of andalusite (a distinctive variety known informally as **chiastolite**). The grain size is coarser than in zone (1), and the rocks might be described as fine-grained phyllites. Cleavage is still preserved, but is even more irregular than in zone (1). In the inner hornfels zone (3) the rocks lose their cleavage and become massive, although the cleavage direction may still be visible as banding. The rocks are even textured hornfelses, and not porphyroblastic.

Figure 3.8 is a drawing of a thin section through a spot in the outer spotted slate zone (1). The section is cut at right angles to the cleavage, as are all thin sections of foliated rocks illustrated in this book. The mineral assemblage of the matrix of the rock is muscovite + chlorite + chloritoid + opaque minerals. The lighter-coloured spot contains muscovite, chloritoid and opaques, but not chlorite. Spots are a characteristic feature of the outer parts of contact aureoles in pelitic rocks. Some of the slates in the outer spotted zone contain biotite, which occurs in rocks of suitable composition, right out to the edge of zone (1). In the inner part of zone (1), porphyroblasts of andalusite and cordierite have been recognized in thin section, although they are not identifiable in the field.

Figure 3.9 shows a cordierite–andalusite slate from the intermediate porphyroblastic slate zone (2). It contains the mineral assemblage biotite + muscovite + cordierite + andalusite + quartz + opaques. The cordierite forms oval crystals with irregular outlines and numerous inclusions of biotite, muscovite and opaques. This is an example of **poikiloblastic texture**, which is the name given when crystals of metamorphic minerals contain inclusions of other minerals. In this case, the inclusions are so numerous that a special name, **sieve texture**, is sometimes used. The biotite and muscovite flakes, and also opaque flakes of graphite, show a strong parallel preferred orientation even within the cordierite crystals. This is relict slaty cleavage surviving from the earlier episode of regional metamorphism and deformation in the Skiddaw slates. The andalusite and cordierite crystals are porphyroblasts, larger crystals in a finer grained matrix. The andalusite crystals are randomly oriented in the plane of slaty cleavage, so that the drawing shows prism sections and basal sections through them. The basal sections at a first glance appear to be twinned crystals, with re-entrant grain boundaries at the corners and a different cleavage direction visible in each segment of the crystal. Examination between crossed polars shows that the crystals are not twinned and their textural features are due to an early stage in the development of the chiastolite habit in the andalusite, which is more fully developed in the next thin section to be described, from the inner hornfels zone (3).

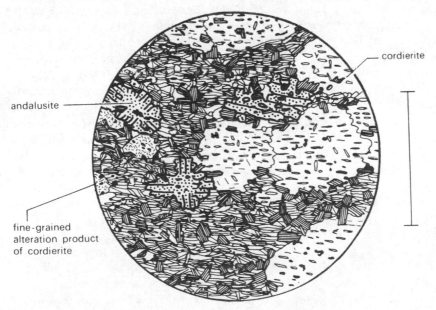

Figure 3.9 Cordierite–andalusite slate from the intermediate zone of the Skiddaw aureole. Scale bar 1 mm.

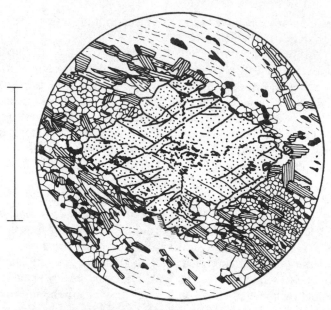

Figure 3.10 Andalusite–cordierite hornfels from the inner zone of the Skiddaw aureole. Scale bar 1 mm.

Figure 3.10 shows a hornfels from the inner zone (3). The mineral assemblage is the same as that in the last thin section, biotite + muscovite + cordierite + andalusite + quartz + opaques. The original slaty cleavage is still visible as preferred orientation of biotite, muscovite and graphite flakes. The cordierite crystals are virtually clear of muscovite and biotite inclusions, but still have graphite flakes and trails of equant minerals indicating the direction of the original slaty cleavage. The andalusite crystal is shown in basal section at right angles to the length of the prism and has a distinctive chiastolite cross of opaque inclusions. The cross has resulted from the mechanism of growth of the andalusite porphyroblast, and it is possible to see how it has arisen by comparing the andalusite crystals in Figures 3.9 and 3.10. In the growing andalusite crystals, boundaries parallel to the {110} prism directions were more stable than the others, and grew more rapidly because of their lower surface energy. In particular, growth at the corners between the {110} faces lagged behind, as can be seen in Figure 3.9. The chiastolite cross formed from graphite inclusions trapped in the re-entrant angles. Thus the arms of the cross in Figure 3.10 represent the traces of the corners as the crystal grew. Because this is a metamorphic rock, and has remained solid throughout metamorphism, the space now occupied by the andalusite crystal must originally have contained muscovite, biotite, quartz, chloritoid and graphite. The growth of the crystal must have involved the introduction of some major chemical constituents, and the removal of

Table 3.1 Analysis of Skiddaw slate from Eastwood *et al.* (1968), compared with an analysis of andalusite.

	Skiddaw slate	Andalusite	Difference
SiO_2	58.41	36.95	− 21.46
TiO_2	1.00		
Al_2O_3	20.25	63.75	+ 43.50
Fe_2O_3	0.63	0.32	
FeO	8.05		− 8.05
MnO	0.07		
MgO	2.02		− 2.02
CaO	0.41		
Na_2O	0.68		
K_2O	2.50		− 2.50
H_2O^+	4.87		− 4.87
H_2O^-	0.46		
P_2O_5	0.23		
BaO	0.04		
C	0.39		
Total	100.01	101.02	

others. What has been gained and lost may be realized by comparing an analysis of Skiddaw slate with an analysis of andalusite (Table 3.1).

Recent research has discovered the link between the evolution of the chiastolite habit in the andalusite and the process of growth of the crystal. Clearly, in the growth process of the andalusite crystal, a large amount of Al_2O_3 was introduced into the volume it now occupies, and a large proportion of SiO_2 was lost. The Skiddaw slates which were metamorphosed lost H_2O during the prograde stages of metamorphism, showing that H_2O was able to diffuse through the rock. The escape of H_2O suggests that the fluid present during metamorphism was H_2O-rich, and carried Al_2O_3 and SiO_2 as ionic complexes in solution out of and into the growing andalusite crystal. However, the presence of graphite in the rock during metamorphism modified the composition of this **metamorphic fluid**. The carbon in the graphite reacted with the fluid, in accordance with the reaction

$$C + 2H_2O = CO_2 + 2H_2$$

The composition of the resultant gas can be calculated, assuming that some H_2O was eliminated by reacting with oxide minerals, and some by reacting with the carbon to form methane gas CH_4. It turns out that under the conditions of metamorphism in the Skiddaw Aureole, the composition of the metamorphic fluid was H_2O 88.0%, CO_2 11.5% and CH_4 0.5%. In the case of SiO_2, it can be shown that the solubility of this chemical component in the metamorphic fluid was significantly lower than it would be in pure H_2O. As a result, SiO_2 was not carried away sufficiently rapidly from the growing crystal face for andalusite to crystallize alone. Instead, andalusite and a proportion of quartz crystallized together (Fig. 3.11). The quartz is present as tube-shaped inclusions, which begin on the arms of the chiastolite cross of inclusions, and run outwards to intersect the prism faces of the crystal approximately at right angles. The cleavage parallel to these quartz tubules is well developed, while the cleavage which cuts across them is relatively undeveloped, and this explains the development of only one of the cleavage directions in each sector of the andalusite crystal. The quartz tubules are clearly visible, and can be picked out as having a lower refractive index by the Becke line test, under high magnification of the microscope (Fig. 3.12).

The hornfels in Figure 3.10 is coarser-grained than the rocks in the outer aureole, and the texture of the quartz crystals can be seen clearly. It shows very well-developed textural equilibrium.

Figure 3.11 Growth of a prism face of an andalusite crystal in the Skiddaw aureole. The face is advancing from left to right, accompanied by an increase in the Al content, and a fall in the Si content, as shown in Table 3.1. Because of the relative insolubility of Al and Si, they are not easily transported away from the growing face by solution in the metamorphic fluid, and therefore Si precipitates as tubes of quartz within the andalusite crystal. The small arrows give an approximate indication of the transport of different elements.

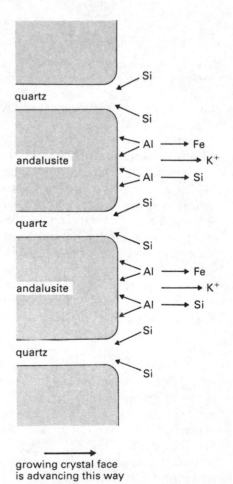

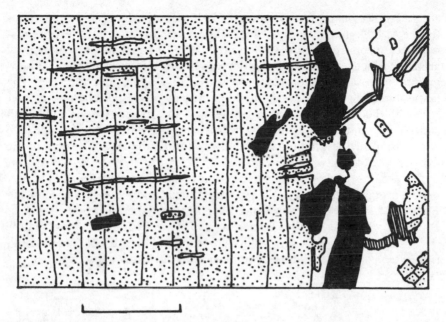

Figure 3.12 The edge of an andalusite crystal from the intermediate zone of the Skiddaw aureole, showing the tubes of quartz present in the andalusite. They do not run right across the drawing because they are slightly oblique to the plane of the thin section. The opaque mineral near the grain boundary is graphite. Scale bar 0.1 mm.

The sequence of mineral reactions which occurred in the Skiddaw Aureole is not easy to determine, because the aureole has been mapped on textural variations in the rocks. Most of the significant changes in the mineral assemblages are seen in the outer spotted zone, but the textural changes show that the effects of heating by the Skiddaw Granite are seen farther away from the contact than the first mineralogical change. The new minerals appearing are first biotite, and shortly afterwards at nearly the same metamorphic grade, cordierite and andalusite. Thus the progressive metamorphic sequence of mineral assemblages appears to be as follows:

chlorite + chloritoid + white mica + quartz + graphite   (Unaltered slates)

biotite + white mica + chlorite + quartz + graphite chloritoid
        + white mica + quartz   (Low-grade assemblages)

cordierite + andalusite + muscovite + biotite + quartz + graphite
(High-grade assemblages)

These assemblages can be represented on AFM diagrams, because they contain quartz and muscovite (Fig. 3.13). The corresponding P–T conditions are also shown on a simple diagram, including equilibrium phase boundaries for the Al_2SiO_5 minerals (Fig. 3.14). While we might think that the progressive metamorphic sequence shown in Figure 3.14 matches a temperature gradient which existed at some time after the intrusion of the Skiddaw granite, more careful consideration shows that this is not the case. The textural relationships between the different minerals, especially in the outer zones of the Skiddaw Aureole, show that the minerals associated with contact metamorphism did not all grow at the same time, relative to the time of intrusion of the granite, and the evolution of the slaty cleavage in the surrounding Skiddaw slates.

The variation of maximum metamorphic temperature with time can be estimated, using an extension of the type of thermal conduction model shown in Figure 3.21. The Skiddaw intrusion is not planar in shape, and a better approximation is a cylinder, with a vertical axis. The roof of the Skiddaw granite intrusion is also not of the simple flat shape shown in Figure 3.21. The distribution of temperature with time around such an intrusion is discussed later.

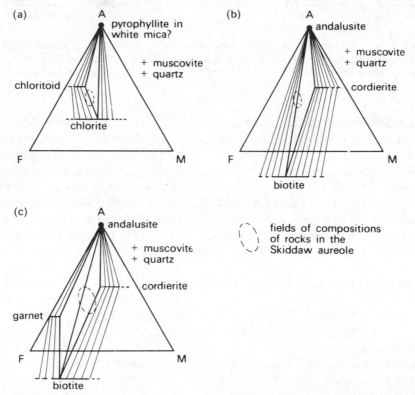

Figure 3.13 AFM diagrams summarizing the mineral assemblages of the Skiddaw aureole. (a) Mineral assemblages of the outer zone (b) Mineral assemblages of the intermediate zone (c) Mineral assemblages of the inner hornfels zone.

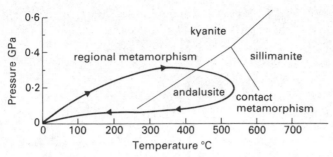

Figure 3.14 P–T t path for rocks in the inner hornfels zone of the Skiddaw aureole, with phase boundaries for the Al_2SiO_5 system.

The model predicts that the rocks near to the intrusion should show simple mineral assemblages, fixed at, or very close to, the maximum temperature achieved in the P–T–t cycle. The rocks farther out might show more complex relationships between metamorphism and textural development. This can be seen in Figure 3.9, where the cleavage in the spotted slates changes near the andalusite porphyroblast. This 'eye structure' of the cleavage is not due to the growth of the andalusite crystal pushing aside the cleavage, as many beginners think (and so did the pioneers of petrology). We can see this by looking at the contacts of the crystal carefully. Slight irregularities in the crystal boundaries of the andalusite show that it replaced quartz and mica crystals slightly less quickly than it advanced along grain boundaries between the matrix crystals. The crystal grew by *replacement*, as we discussed earlier in connection with the chemical changes involved in crystal growth. The 'eye structure' has arisen because the deformation which caused the slaty cleavage continued to be developed after andalusite growth (Fig. 3.15). Eye structure of this kind is also often seen in regional metamorphic rocks, and arises in a similar way (see Chapter 7).

We have arrived at a more detailed picture of the evolution of the contact aureole at Skiddaw than our initial simple idea of regional metamorphism, followed by contact metamorphism which overprinted the regional metamorphic minerals and their associated structures. It is clear that the Skiddaw granite is not **post-tectonic** but **syn-tectonic**, and was intruded near the end of the major tectonic episode. This conclusion is confirmed by study of radiometric dates on the Skiddaw granodiorite and on the Skiddaw slates (Soper 1987). The tectonic and metamorphic history of the Skiddaw Aureole can be tabulated as follows (Table 3.2).

This summary of the history of the Skiddaw contact aureole brings out the nature of both the regional and contact metamorphism. They were not brief events, fixing high temperature mineral assemblages at a particular instant, but both went on over considerable periods of time. The regional metamorphic P–T–t cycle was longer than the contact metamorphic one, so that the contact metamorphism occurred during the retrograde stages of the regional cycle.

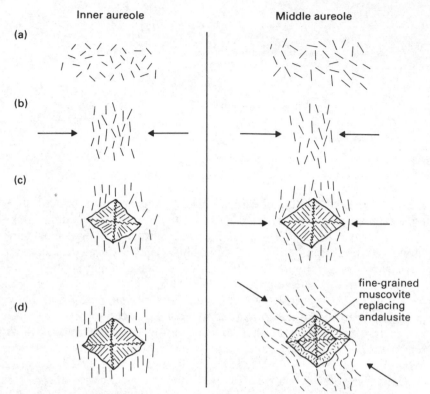

Inner aureole   Middle aureole

(a)

(b)

(c)

(d)

fine-grained
muscovite
replacing
andalusite

Figure 3.15 Sequence of growth of andalusite porphyroblasts in the Skiddaw aureole. (a) before deformation and metamorphism, (b) development of slaty cleavage, during Caledonian deformation, (c) growth of andalusite, after intrusion, when Caledonian deformation had resumed, (d) inner aureole – strong hornfels protects andalusite and surrounding groundmass from further deformation and metamorphism. Outer aureole – matrix is deformed around porphyroblast, retrograde metamorphism of outer parts of andalusite to fine-grained muscovite.

The origin of the spots in the outer spotted zone, shown in Figure 3.9, is of interest at this point in the discussion. Some of the spots are not oval but angular in outline, suggesting that they were formed by the replacement of andalusite rather than cordierite. This speculation may be confirmed by studying other contact aureoles in pelitic rocks. In the contact aureole of the Zhoukoudian Granite, near Beijing, China, andalusite crystals in the inter-mediate zone show *partial* replacement by hydrous minerals. These are chloritoid (in radiating clusters) and fine-grained muscovite. This shows that similar replacement of andalusite by cordierite and muscovite is plausible in the Skiddaw Aureole as well. The time–temperature curves shown in Figure 3.2 indicate that the maximum temperature (i.e. the metamorphic peak) of contact metamorphism occurred later in the outer aureole than it did in the inner aureole. The replacement of andalusite by chloritoid and muscovite

Table 3.2 Tectonic and metamorphic history of the Skiddaw contact aureole.

Early events	Deposition of Skiddaw slates	
Caledonian orogenesis	Folding of Skiddaw slates Initiation of slaty cleavage along axial planes of folds	Ordovician
	Metamorphism of clay minerals to chlorite, white mica and chloritoid	Silurian
	Intrusion of Skiddaw granite Onset of growth of contact metamorphic minerals in contact aureole	Devonian
	Growth of minerals in spotted zones	
	Intensification of slaty cleavage, leading to 'eye structure' round porphyroblasts	
	Replacement of high temperature contact metamorphic minerals by lower temperature minerals (resemble the earlier Caledonian regional metamorphic minerals)	

occurred during later regional metamorphism. If the Skiddaw granite was intruded after the metamorphic peak of regional metamorphism, but before the rocks had cooled to near-surface temperatures, the spots can be explained, not by a separate later episode of regional metamorphism, but as having occurred during the later stages of the regional metamorphic P–T–t cycle.

The Beinn an Dubhaich aureole, Isle of Skye, Scotland

Contact metamorphism of carbonate-rich sediments introduces a new factor in metamorphism. The example to be described here is the contact aureole surrounding the Beinn an Dubhaich granite, Isle of Skye, Scotland (Fig. 3.16).

The country rock surrounding the granite is the Suardal Dolomite Formation, of Cambrian age. It is a dolomite rock or dolostone containing oval nodules of chert, often concentrated in layers parallel to the bedding. Close to the contact of the granite, the dolomite of the country rock has a changed mineral assemblage due to the contact metamorphism. The $MgCO_3$ component in the dolomite has broken down by the following metamorphic reaction:

$$CaMg(CO_3)_2 = CaCO_3 + MgO + CO_2$$

The CO_2, carbon dioxide gas, has escaped from the rock. This reaction was described long ago by the pioneer metamorphic geologist Alfred Harker (1932) who named it **dedolomitization**. Periclase is not stable in rocks near the surface of the earth through which ground water circulates, and has become altered by hydration to brucite $(Mg(OH)_2)$. The reader may wonder why the equivalent reaction in calcite has not occurred in the Beinn an Dubhaich Aureole. It is:

$$CaCO_3 = CaO + CO_2$$

This reaction is used in the manufacture of quicklime, and is often performed in the laboratory by heating powdered calcite in a crucible over a gas burner. The reason the reaction is not seen in contact metamorphic rocks is that a small increase in pressure above atmospheric pressure inhibits the escape of carbon dioxide, even at the high temperatures of contact metamorphism.

The chert nodules introduce an additional chemical component, SiO_2, into the dolomitic country rocks. As the contact of the intrusion is approached,

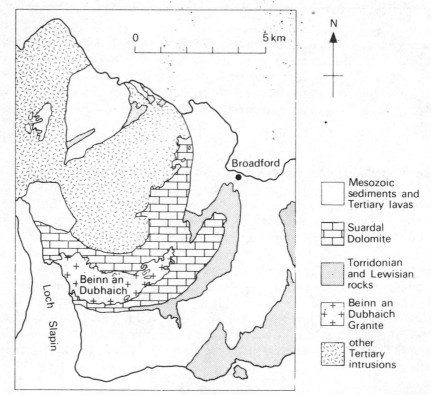

Figure 3.16 Geological map of part of the Tertiary igneous centre of the Isle of Skye, Scotland, showing the Beinn an Dubhaich Granite, from Craig 1983.

reaction rims appear between the nodules and the dolomite, indicating that the quartz of the chert has reacted at the surfaces of the nodules with some component in the surrounding carbonate rock. The sequence of new minerals appearing in the reaction rims with increasing metamorphic grade is – talc ($Mg_3Si_4O_{10}(OH)_2$), tremolite ($Ca_2Mg_5Si_8O_{22}(OH)_2$), diopside ($CaMgSi_2O_6$), forsterite (Mg_2SiO_4), periclase (MgO) and wollastonite ($CaSiO_3$).

The reaction to form talc has occurred because dolomite and quartz ceased to coexist in equilibrium together as the temperature rose due to heating by the intrusion. Reaction began at the surfaces of the nodules, where quartz and dolomite crystals were in contact, or very close to one another. SiO_2 diffused out of the chert nodules, and MgO out of the surrounding dolomite rock, to crystallize in the widening reaction rims. Closer to the contact, at higher temperatures, CaO began to diffuse as well. The structure of the reaction rims was controlled by the effectiveness of diffusion.

The talc-forming reaction is more complex than the simple reactions we have considered so far. It is

(1) $3CaMg(CO_3)_2 + 4SiO_2 + H_2O = Mg_3Si_4O_{10}(OH)_2 + 3CaCO_3 + 3CO_2$

For the moment we should note that this reaction involves the loss of CO_2, but the introduction of H_2O. Presumably the water was introduced to the rock as ground water at the time of intrusion, which was in Cenozoic times, long after the deposition of the sediment. The carbon dioxide escaped from the rock. The reaction is said to be a **decarbonation** reaction, but also a **hydration** reaction.

As the metamorphic grade increases towards the contact of the intrusion, tremolite joins talc in the reaction rims. It is formed by the reaction:

(2) $5CaMg(CO_3)_2 + 8SiO_2 + H_2O = Ca_2Mg_5Si_8O_{22}(OH)_2 + 3CaCO_3 +$
       $7CO_2$

Like the talc-forming reaction, this is a hydration and decarbonation reaction.

The subsequent reactions to form diopside, forsterite, wollastonite and periclase are as follows:

(3) $CaMg(CO_3)_2 + 2SiO_2 + CaMgSi_2O_6 + 2CO_2$

(4) $CaMgSiO_2O_6 + 3CaMg(CO_3)_2 = 2Mg_2SiO_4 + 4CaCO_3 + CO_2$

(5) $CaCO_3 + SiO_2 = CaSiO_3 + CO_2$

(6) $CaMg(CO_3)_2 = MgO + CaCO_3 + CO_2$

78

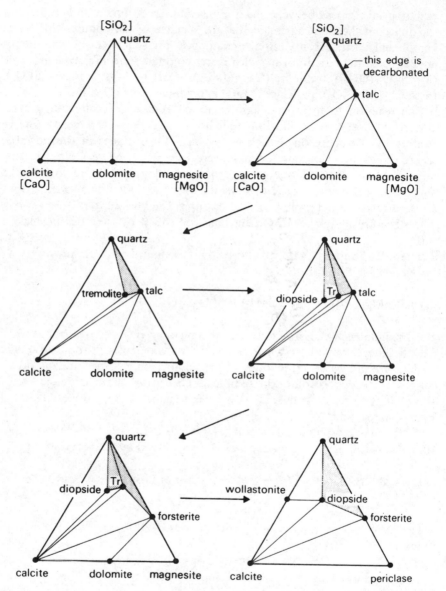

Figure 3.17 A sequence of triangular diagrams to illustrate the progressive decarbonation of siliceous dolomite in the Beinn an Dubhaich aureole. Shaded parts of the triangles represent assemblages which are completely decarbonated.

All these are decarbonation reactions, involving the loss of carbon dioxide from the rock. All the minerals appear in reaction rims around the chert nodules (which change to crystalline quartz in the aureole), sometimes as alternating layers except for periclase, which appears as isolated porphyroblasts in the dolomite part of the rock.

The change in the equilibrium mineral assemblages with increasing metamorphic grade are summarized in Figure 3.17. The chemical composition of the dolomite rock with its included chert nodules can be described in terms of just four chemical components CaO, MgO, CO_2 and SiO_2. All possible rock compositions can therefore be plotted inside a tetrahedron with these four components at its corners. The triangles in Figure 3.17 are projections of points representing rock compositions onto the SiO_2–MgO–CaO face. The presence of the hydrous minerals talc and tremolite at lower metamorphic grades in the aureole presents a complication, which is overcome by neglecting the H_2O content of both rocks and minerals in plotting the diagram.

From the equations for the reactions, and from Figure 3.17, it can be seen that with increasing metamorphic grade, all the chemical reactions involve the loss of CO_2 from the rock. Thus the metamorphic sequence may be described as displaying progressive **decarbonation** with increasing metamorphic grade. The carbon dioxide escaped from the rocks during metamorphism, similar to water from the rocks in the Skiddaw aureole.

The minerals formed in the sequence of reactions in the Beinn an Dubhaich aureole have fixed chemical compositions, and the conditions under which they are stable have been well established by experiment and by thermodynamic calculation. The temperatures at which all the components in each of the equations (1) to (6) co-exist in equilibrium at a pressure of 101.3 kPa (1 atmosphere) are included in Table 3.3. The experimentally determined sequence of reactions with increasing temperature at low pressure therefore matches the naturally occurring sequence at Beinn an

Table 3.3   Reactions in the system CaO − MgO − SiO_2 − CO_2 at a pressure of 101.3 kPa.

Reaction no.	Reaction type*	Reaction temperature (°C)
(1)	H + DC	185
(2)	H + DC	190
(3)	DC	210
(7)	DC + DH	240
(8)	DC + DH	280
(9)	DC + DH	305
(4)	DC	310
(5)	DC	310
(6)	DC	410

*Reaction types: H – hydration, DC – decarbonation, DH – dehydration

Dubhaich, suggesting that the contact aureole formed at a relatively shallow depth.

There is a complication, however, in regard to pressure. We saw how reactions (1) and (2) required the introduction of H_2O into the rock as well as the driving-out of CO_2. The sequence of triangles in Figure 3.17 not only shows the consequences of reactions (1) to (6), which introduce new minerals with increasing metamorphic grade, it also shows reactions in which the hydrous low temperature minerals are replaced by anhydrous minerals at higher temperatures i.e. **dehydration reactions**:

(7)　$Ca_2Mg_5Si_8O_{22}(OH)_2 + 3CaCO_3 + SiO_2 = 5CaMgSi_2O_6 + 3CO_2 + H_2O$

(8)　$Ca_2Mg_5Si_8O_{22}(OH)_2 + 3CaCO_3 = 4CaMgSi_2O_6 + CaMg(CO_3)_2$
　　　$+ H_2O$

(9)　$Ca_2Mg_5Si_8O_{22}(OH)_2 + 11CaMg(CO_3)_2 = 8Mg_2SiO_4 + 13CaCO_3$
　　　$+ 9CO_2 + H_2O$

Since we now have quite a number of possible reactions, let us consider them in order of increasing temperature of reaction at 101.3 kPa.

The different reactions listed in Table 3.3 will be affected by the nature of the fluid present at the time of metamorphism. Reactions (1) and (2) could not occur if H_2O was not present to form talc and tremolite respectively, since both contain essential H_2O in their crystal structure. All reactions in which H_2O or CO_2 appear on one side or the other of the reaction equation will be influenced by the composition of the fluid concerned. If the fluid is mostly H_2O, hydration reactions will tend to occur at lower temperatures, dehydration reactions at higher temperatures.

Since the rocks concerned in the Beinn an Dubhaich Aureole are rich in CO_2 and relatively poor in H_2O, the reaction temperatures in Table 3.3 are those which would be found if the fluid were rich in CO_2, but contained an appreciable proportion of H_2O.

Figure 3.18 shows a set of univariant reaction curves for reaction (5) on a P–T graph. X_{CO_2} is the proportion of CO_2 in the fluid. To avoid the complications introduced by the possibility of reaction of the rest of the fluid present, the non-reactive gas argon has been used as the dilutant. The curve labelled 0.5 therefore represents the univariant curve along which calcite, quartz, wollastonite and argon + CO_2 mixture co-exist in equilibrium. The different curves show clearly that the temperature at which calcite and quartz will react together to produce wollastonite varies with fluid composition, at 0.2 GPa (2 kbar) for example from below 400°C to 700°C. The lower temperature applies when the gas is virtually all argon, the higher when it is all CO_2.

Figure 3.18 shows univariant reaction curves at 0.1 GPa pressure (1 kbar)

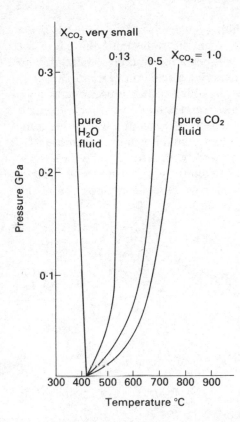

Figure 3.18 P–T plot of univariant curve for reaction (5) in the text, for different proportions of CO_2 in the metamorphic fluid. The curve on the right is for a pure CO_2 fluid, that on the left for a fluid in which the proportion of CO_2 is effectively zero.

for fluids which are a mixture of CO_2 and H_2O. An additional reaction is plotted:

(10) $4Ca_2Al_3SI_3O_{12}(OH) + SiO_2 = 5CaAl_2Si_2O_8 + Ca_3Al_2O_{12} + 2H_2O$

zoisite + quartz = anorthite + grossular + H_2O

These reactions are of different types – (5) is a decarbonation reaction (DC), (10) a dehydration reaction (DH), (1) a decarbonation + hydration reaction (DC + H) and (7) a decarbonation + dehydration reaction (DC + DH). With a starting assemblage tremolite + zoisite + dolomite + - calcite + quartz, the sequence of new minerals appearing with increasing temperature, if the fluid composition were kept constant at X_{CO_2} 0.2, X_{H_2O} 0.8, would be diopside, talc, wollastonite, anorthite; at X_{CO_2} 0.8, X_{H_2O} 0.2 it would be anorthite, diopside, wollastonite.

What happens in nature when rocks containing both CO_2 and H_2O as essential components in minerals, (in contrast to fluids in cracks, fluid inclusions or adsorbed on grain boundaries), are heated? Consider a simple three-phase assemblage zoisite + dolomite + quartz, and consider that it is initially in contact with a fluid with X_{CO_2} 0.3, X_{H_2O} 0.7 (Fig. 3.19). On heating, the fluid composition will remain the same until the curve for reaction (1) is reached at 520°C. Talc will begin to crystallize, and dolomite and quartz cease to be stable together. Since reaction (1) is a decarbonation + hydration reaction, CO_2 will be added to the intergranular fluid, and H_2O will be extracted from it. Unless fluid is readily available from an outside source, (for example by circulation of ground water), the composition of the metamorphic fluid will change, and it will become more CO_2 rich. The temperature will continue to increase, and for a while the assemblage talc + zoisite + dolomite + calcite + quartz will continue to be in equilibrium with vapour. Because the vapour now varies in composition as the

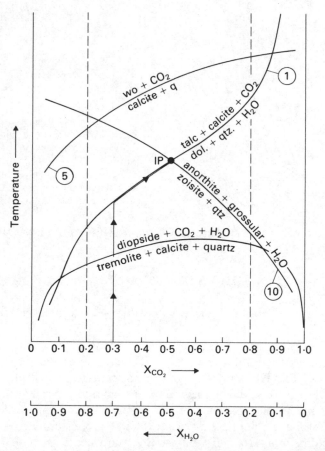

Figure 3.19 Univariant reaction curves for reactions (1), (5) and (10) and also for the reaction tremolite + calcite + quartz = diopside + CO_2 + H_2O, for different compositions of metamorphic fluid, from pure H_2O on the left, to pure CO_2 on the right. The curve with arrows shows how the composition of a metamorphic fluid with an initial composition CO_2 30%, H_2O 70% would change with increasing temperature, if its composition were internally buffered. IP – invariant point at which composition of fluid would be fixed, until one of the reactants is used up.

temperature rises, the system has gained a degree of freedom, and the number of phases which can co-exist has correspondingly increased by 1.

What happens next will depend on the relative proportions of the mineral phases. If all the dolomite or quartz is used up by the reaction, the temperature will rise above the curve corresponding to reaction (1), and the fluid composition will remain at the last value it held when the curve was left. If the temperature were to rise to 560°C, and quartz was still present, zoisite would begin to break down by reaction (10). This reaction would release H_2O, and halt the change in the fluid to more CO_2-rich compositions. The rise of temperature would be checked until one of the minerals dolomite, zoisite or quartz was used up. Meanwhile, the fluid composition would remain fixed at X_{CO_2} 0.5, X_{H_2O} 0.5. In this condition, the fluid composition is said to be '**buffered**' at a fixed value by the presence of the minerals participating in the reactions.

Since the minerals which are causing the buffering, dolomite and zoisite, are present within the rock undergoing metamorphism, the relative proportions of CO_2 and H_2O, X_{CO_2} and X_{H_2O}, are said to be **internally buffered**. Where the composition of the fluid present during metamorphism is controlled by metamorphic reactions occurring outside the rock, (for example in an adjacent layer, which is releasing fluid due to metamorphic reactions), the composition of the fluid is said to be **externally buffered**. In the carbonate rocks of the Beinn an Dubhaich aureole, internal buffering would explain the high X_{CO_2} indicated by the mineral assemblages in the inner parts of the aureole. By contrast, in the Skiddaw aureole, the widespread presence of H_2O bearing (hydrous) minerals indicates that X_{H_2O} should be high.

Alfred Harker (1932) recognized that the sequences of rocks seen in contact metamorphic aureoles provided an insight into metamorphic processes which also control the rocks of progressive metamorphic sequences in regional metamorphic rocks. This insight is still valuable, and points to carry forward from the discussion of contact metamorphic rocks are as follows:

(1) The most important control on the mineral assemblages and textures in progressive metamorphic sequences is the temperature at the metamorphic peak.

(2) This metamorphic peak temperature is achieved at different times in different parts of the sequence.

(3) The composition of the metamorphic fluid also has a profound influence on both the mineral assemblages and textures.

(4) This composition is frequently internally buffered by the mineral assemblages in the rocks.

Evidence for the origin of the metamorphic fluid in the Beinn an Dubhaich aureole has been obtained from the study of the stable isotopes of oxygen and hydrogen. The ratios have been measured in H_2O extracted from minerals which crystallized during contact metamorphism.

Compositions of fluids in contact metamorphism

Oxygen and hydrogen are present in water as isotopes, i.e. atoms with the same numbers of electrons surrounding their nuclei, but different numbers of protons + neutrons in their nuclei, so that their atomic masses are different, although their chemical properties are virtually identical. The ratios of the different isotopes of oxygen and hydrogen differ very slightly in water from various sources. In the case of natural rain water or snow, this variation is due to natural fractionation of isotopes occurring during the continuing hydrological cycle. For water locked into the crystals of metamorphic rocks, there was also isotopic fractionation at the time of metamorphism. The proportions of the different isotopes of oxygen and hydrogen in sea water are as follows: ^{16}O 99.759%, ^{17}O 0.037%, ^{18}O 0.204%; ^1H 99.985%, ^2H (D) 0.015%. ^3H is present in very small amounts in sea water and rain water, but is entirely an artificial product of hydrogen bomb testing, and so is not present in metamorphic rocks.

The variations in isotopic ratios are small. Two ratios are commonly measured – that of ^{18}O to ^{16}O, and that of deuterium (^2H, or D) to hydrogen (^1H). The ratios of these isotopes, measured by means of a mass spectrometer, are compared with the standard ratios of sea water, because sea water is well mixed and constitutes the Earth's reservoir of H_2O. The standard is called Standard Mean Ocean Water, which is abbreviated to **SMOW**.

The relative concentration $^{18}O/^{16}O$ or D/H in a particular sample is expressed as the per thousand concentration δ (delta), defined as follows: for ^{18}O

$$R_{sample} = {}^{18}O/^{16}O \text{ in sample}$$

$$R_{SMOW} = {}^{18}O/^{16}O \text{ in SMOW}$$

$$\delta^{18}O = (R_{sample}/R_{SMOW} - 1).1000$$

for D

$$R_{sample} = D/H \text{ in sample}$$

$$R_{SMOW} = D/H \text{ in SMOW}$$

$$\delta D = (R_{sample}/R_{SMOW} - 1).1000$$

Positive values of δ show that the sample is enriched in deuterium or ^{18}O compared with sea water, for negative values that it is depleted in these isotopes.

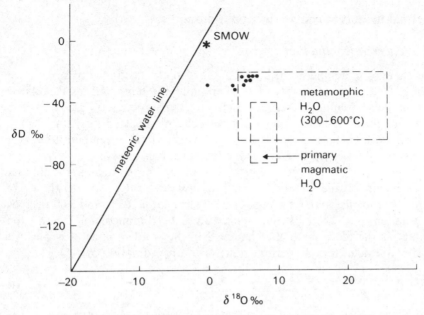

Points are possible modern metamorphic waters from California

Figure 3.20 Oxygen and hydrogen isotopic variation in meteoric waters (line), primary magmatic waters and metamorphic waters (fields). From Taylor (1974).

Meteoric water, that is rain water and snow, has a distinctive range of hydrogen and oxygen isotope ratios, related to the mean annual temperature of the climate and therefore the geographical latitude of the point where it falls. On a diagram plotting δD against $\delta^{18}O$, these meteoric waters define a straight line (Fig. 3.20). This line is called the **meteoric water line**. Primary magmatic water, which has come from the source region of magmas in the upper mantle, has a different small range of isotopic compositions, also shown in Figure 3.20. The water from metamorphic rocks, which is the residue left after partial dehydration shows a wider range of isotopic compositions. These differences in isotopic composition permit the origin of the water, trapped in minerals during metamorphism, to be determined in many cases (Taylor 1974, Taylor & Forester 1971). The values of $\delta^{18}O$ have been more widely measured than those of δD, and can be used on their own to find the origin of metamorphic water in suitable cases. The measurements show that the minerals in the Beinn an Dubhaich aureole absorbed circulating ground water during metamorphism, suggesting that this aureole formed at a shallow level, and heat was transferred by convection rather than conduction, as discussed in the following section.

Thermal models of contact metamorphism

(i)　Thermal conduction models

The changes of temperature in the country rocks around intrusions have been modelled by a number of workers, and the temperatures calculated from models based upon thermal conduction across contacts agree well with the temperatures measured in contact aureoles by study of the metamorphic mineral asssemblages. Figure 3.2 shows distributions of temperatures at a succession of times (t_1, t_2, t_3) after the time of intrusion (t_0).

The maximum temperature always occurs immediately against the contact of the intrusion, the temperature falling away with distance. In the early stages after intrusion (t_1), the temperature gradient is steep, and as time goes on it flattens out. The maximum temperature T_{max}, immediately against the contact, is therefore the highest reached in the whole contact aureole. An empirical 'rule of thumb' permits quite accurate estimation of T_{max}.

$$T_{max} = 2.(T_m - T_c)/3 + T_c \qquad \text{(i)}$$

where T_m is the temperature of the magma when it is intruded and T_c the temperature of the country rocks immediately before intrusion. Thus for an example where the temperature of the magma (T_m) is 1000°C (a reasonable value for a magma of basaltic composition) and the maximum temperature of the country rock 100°C, T_{max} may be estimated by substitution in the equation as follows:

$$T_{max} = 2.(1000 - 100)/3 + 100°C$$

giving a value for T_{max} of 700°C.

The effectiveness of this purely empirical equation can be compared with

Table 3.4　Maximum contact temperatures T_{max} computed for intrusion of magma at temperature T_m into dry country rocks at temperature T_c, by thermal conduction calculations (T_{max} [1]) (Turner 1981) and by using empirical equation (i) (T_{max} [2]).

		Temperature °C			
Igneous rock	Country rock	T_m	T_c	T_{max} [1]	T_{max} [2]
Granodiorite	Shale	800	100	685	567
Granodiorite	Shale	800	200	710	600
Granodiorite	Sandstone	800	100	~ 500	567
Gabbro	Shale	1000	0	750	667
Gabbro	Sandstone	1000	0	~ 460	667
Gabbro	Sandstone	1000	100	~ 500	700

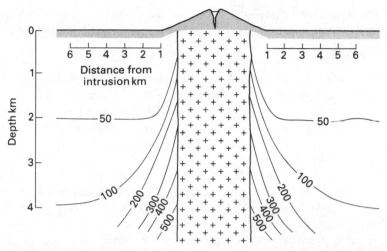

Figure 3.21 Distribution of isotherms round a cylindrical intrusion with a diameter of 4 km, 100 000 years after the intrusion. (Note that the vertical scale is exaggerated).

values calculated using estimates of the thermal conductivities and thermal capacities of the igneous rocks and country rocks, and the thermal conduction equation.

Remember that these figures are based on modelling, but that they agree reasonably well with the results of temperature estimation using mineral assemblages in hornfelses.

The maximum temperature at the contact of an intrusion is independent of the size of the intrusion, and also of its shape, (provided that the contact is not irregular on a scale of a few metres). The temperature *distribution*, on the other hand, is very dependent on the size of the intrusion. If the intrusion is large, heating will occur farther out in the country rocks than it will for a small intrusion. It is impossible to deal with all cases here, so only one will be discussed, that of a vertical cylindrical intrusion of granitoid magma, 4 km in diameter, with an initial temperature of 800°C, which is approximately the size and magma temperature of the Skiddaw Granite of Cumbria. Figure 3.21 shows a cross section through the intrusion and the temperature distribution 100 000 years after intrusion.

Before the significant features displayed by this model are discussed, the assumptions made in constructing the model should be summarized. They are:

(1) All heat transport in solid igneous rocks and country rocks was by thermal conduction.
(2) The intrusion formed instantaneously, and the intruded magma remained in it, crystallizing and cooling, without being withdrawn or recharged.

(3)   The magma was intruded into continental crust of uniform thermal conductivity, which had previously arrived at a steady-state geothermal distribution.

It can be seen that the vertical batholith produces large thermal gradients in a horizontal direction, near the contacts. Since the metamorphic reactions of contact metamorphism are frequently dehydration reactions, with steep slopes in P–T space, it can be assumed that isotherms on the cross section Figure 3.21 are approximately parallel to **isograds** defined by the incoming of contact metamorphic minerals, such as andalusite and cordierite. It follows that the metamorphic aureole surrounding the intrusion will increase in width at deeper levels in the Earth's crust. This conclusion may be checked by study of eroded contact aureoles, and has been found to be correct in a number of cases.

The predictions of the model concerning the change in temperature with time in the contact aureole are also interesting. Figure 3.22 shows the change of temperature with time as a number of curves, one for temperature immediately against the intrusion, and others for temperatures at 1 km, 2 km from the intrusion, and so on. These show two things:

(1)   the maximum temperature attained, and therefore the metamorphic grade, declines with distance away from the intrusion.
(2)   the maximum temperature is achieved at later and later times with distance away from the intrusion.

It can be said that a 'wave' of high temperature moves away from the intrusion, diminishing in height as it does so. This observation fits in well with the relationships of the andalusite porphyroblasts with the regional

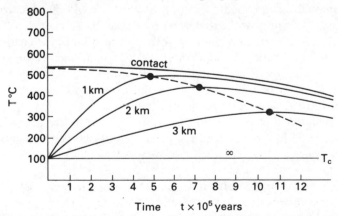

Figure 3.22   Temperature–time (T–t) curves for rocks at different distances from the intrusion, and at 4 km depth, for the intrusion shown in Figure 3.21. Dotted line, times of maximum temperature (Turner 1981).

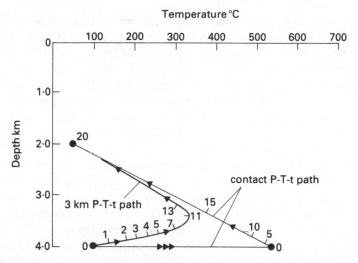

Figure 3.23 Depth–temperature–time (d–T–t) curve for the intrusion shown in Figure 3.21, with the additional assumption that there was uplift after intrusion at the rate of 2 mm per year. Both rocks were initially buried at 4 km, one very close to the contact, and one 3 km away. The numbers on the curve give the depth and temperatures at intervals of 100 000 years after the intrusion (i.e. 5 = 500 000 years after intrusion).

metamorphic cleavage in the Skiddaw slates. If it is the case that the cleavage of the Skiddaw slates evolved progressively over a period of time, rather than in two separate events before and after the intrusion of the Skiddaw Granite, then the intensification of the cleavage leading to its 'wrapping around' the andalusite porphyroblasts in the middle zone of the aureole can be understood.

P–T–t paths for rocks in the aureole will not show a variation in pressure, because in the simple model shown in Figures 3.21 and 3.22, the depth of burial of all rocks in the aureole remains the same throughout the sequence of events. More useful and understandable P–T–t paths can be prepared by developing a model in which the intrusion of the granitoid is followed by a steady rate of uplift of 2 mm per year. This is quite a plausible possibility, because the high heat flow into the upper lithosphere which often accompanies magmatic intrusion, could cause expansion of the lower lithosphere, and uplift of the surface.

Figure 3.23 is a temperature/depth plot for two rocks in the contact aureole – one immediately adjacent to the contact of the intrusion, the other for a rock 3 km from the contact.

The rock at the contact heats instantaneously to the value for T_{max} at a depth of 4 km. The subsequent cooling is slow, the rock taking 1 million years to cool from 530°C to 500°C. After the initial period of cooling over a relatively small temperature interval, cooling becomes more rapid so that the temperature has fallen below 100°C after 2 million years. The rock 3 km from the intrusion reaches a lower maximum temperature of 350°C, reaching this maximum about 1 million years after intrusion. The temperature remains around the maximum for about 500 000 years, and then cooling becomes more rapid.

Although there are many assumptions involved in this model, it points to a number of significant features of contact metamorphism where heat transfer is dominated by conduction. The shape of the P–T–t paths suggests that contact metamorphic rocks are likely to have remained near their maximum temperature for a substantial period of time in their history, so that they are likely to contain equilibrium mineral assemblages which crystallized close to their maximum temperature. This fits with textural studies of hornfelses, which display equilibrium textural features. It also explains why contact metamorphic rocks often define a metamorphic facies well, for example the contact metamorphic assemblages described by Goldschmidt (1911) from the Oslo region of Norway. The variation in the timing of attainment of maximum temperature is a significant feature of contact aureoles. The variation might be detectable by studying radiometric dates from a contact aureole. The progressive metamorphic sequences obtained by studying the mineral assemblages of the contact aureole would not represent thermal gradients existing at any one time in the evolution of the contact aureole.

(ii) Thermal convection models

The kind of model described above does not apply in all cases, because igneous rocks are frequently intruded high in the lithosphere in rocks which contain pore fluid, and fluid in fractures. If the intrusion lies at a level in the crust where ground water can circulate, the flow of hot fluid, combined with the high thermal capacity of water, will make fluid flow a far more effective mechanism in removing heat than conduction through the rocks or water. **Thermal convection** rather than **thermal conduction** will be the controlling mechanism.

It is possible to create models for this kind of heat transfer, but they involve estimation of many properties of rocks, porosity, permeability, thermal capacity, for example. A quantitative treatment will not be attempted here, but a simple qualitative model will be introduced for comparison with the conductive model discussed above.

Figure 3.24 illustrates the flow of heat and fluid across a vertical contact of an intrusion, a short time after the magma has been intruded. The intrusion has already developed a thin chilled margin of solid igneous rock, and heat is transferred through this by conduction. The large surface area of grains in the permeable country rock transfers this heat very efficiently to the pore water close to the contact, and this heated water becomes less dense and rises along the contact. Cool water is drawn in from the country rock to replace the rising hot water. The temperature gradient will be large across the

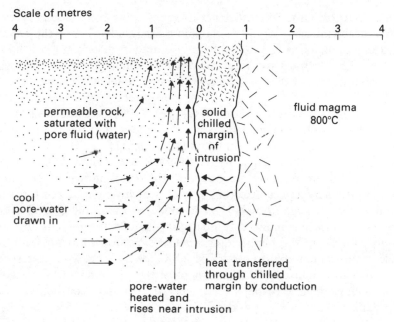

Scale of metres

permeable rock,
saturated with
pore fluid (water)

solid
chilled
margin
of
intrusion

fluid magma
800°C

cool
pore-water
drawn in

heat transferred
through chilled
margin by conduction

pore-water
heated and
rises near intrusion

Figure 3.24 Flow of heat across the contact of an intrusion, and flow of ground water near the intrusion, where the country rocks are permeable. Straight arrows represent fluid flow, wavy arrows heat flow. Compare Figure 3.2.

boundary between the rising hot water and the surrounding cold water. A convection flow system is set up in the ground water, which rapidly carries heat away from the contact of the intrusion. Whereas, in the conduction case (Fig. 3.22), the temperature gradients tended to become less steep with time, in the convective case temperature gradients will tend to remain high until the temperature of the intrusion has dropped to a level at which the convection system will cease to operate. At shallow levels the permeable layer can be recharged by ground water which has fallen as rain or snow (meteoric water), and the heated water may appear at the surface as hot springs.

Figure 3.25 shows an intrusion of similar size and magma temperature to that of Figure 3.21, but this time it has intruded into permeable rocks charged with pore water. The small arrows show the direction and velocity of flow of the pore water at a time shortly after the intrusion occurred, although there has been time for the convective circulation system to become established.

The temperature distribution shows a number of significant differences from that in the conduction model. The contact aureole would remain narrow, and would be uniform in width alongside the contacts. It could actually become wider upwards, as the hot ground water spreads out above cold ground near the ground surface. This could give rise to local inverted geothermal gradients. The efficiency of convective heat transport would mean that the intrusion would cool much more rapidly than in the conductive case, although it is very difficult to give numerical estimates of the time involved. The outward travelling wave of maximum temperatures, which was such a feature of the conductive transport model, will not occur in this model.

Figure 3.25 shows the intrusion 100 000 years after intrusion has occurred. Because the granitoid magma has solidified, the pore water may enter the solid igneous rock of the intrusion along fractures. The fracturing of the hot igneous rock is actually helped by the influx of cooler fluid. The convective circulation system can therefore pass into the intrusion itself. The country rocks immediately adjacent to the contact, which remained hot for a long period in the conduction model, might undergo quite quick cooling from their temperature maximum, as those rocks pass from the uprising column of hot pore fluid into the surrounding area of slow down-flowing cold fluid. While, with the convection model, it is not possible to work out specific P–T–t paths for individual rocks, it is clear that there could be periods of rapid cooling, and also perhaps heating, as the pattern of convective circulation changes over time.

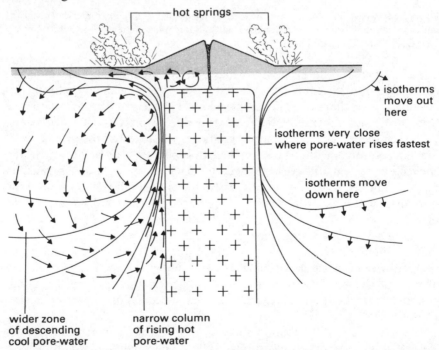

Figure 3.25 Pattern of isotherms and water flow for a cylindrical intrusion 4 km in diameter, where heat is mainly transferred by convection in the ground water of the country rocks. The diagram is qualitative, but is of approximately the same scale as Figure 3.21, including the vertical exaggeration.

This type of model may seem very speculative, but it is powerfully supported by the evidence given by the geochemistry of the stable isotopes of oxygen and hydrogen in the H_2O in metamorphic rocks when the origin of the water present during contact metamorphism is being considered.

4 Dynamic metamorphism

In the upper parts of the Earth's lithosphere, where the rocks are relatively cool, most of the deformation takes place by fracturing, giving rise to joints, faults and thrusts. Because the temperatures are relatively low, this deformation may not be accompanied by metamorphic reactions, involving the breakdown of one mineral species and the appearance of new mineral species. *Mechanical* processes are dominant. However, within and close to fractures, mechanical processes frequently operate on the grain-size scale, producing very extensive modifications of rock fabrics. There may be some metamorphic reactions, but the mechanical processes are the dominant ones.

Dynamic metamorphism is the term used to describe these processes in which mechanical deformation of all sorts dominates over metamorphic reaction in giving rise to the metamorphic rocks (Brodie & Rutter 1985). As with contact metamorphism, the areas affected by dynamic metamorphism are relatively small. There is a dynamic element to a great deal of regional metamorphism, which is therefore sometimes called 'regional dynamo-thermal metamorphism', and the boundary conditions between dynamic metamorphism and regional metamorphism are not clearly defined. At present, it is probably best to adopt a practical, field-based approach to the definition (Chapter 1), and to define dynamic metamorphic rocks as those which are associated with intense deformation in or near fracture zones. The fracture zones concerned can usually be identified by field geological studies. A feature of dynamic metamorphic rocks is that, because mechanical processes of deformation and fragmentation are the distinctive feature of their metamorphism, textures are more important, and lists of minerals present less important, than in contact and regional metamorphic rocks. Many dynamic metamorphic rocks do not contain equilibrium assemblages of minerals. Some do, and these are of interest because conditions of temperature, pressure and fluid composition during metamorphism, and therefore deformation, can be estimated from them. But many dynamic metamorphic rocks with very interesting and informative textures contain only one type of mineral, such as quartz or calcite. These are revealing, not because of the P–T conditions of metamorphism, but because the experimentally determined mechanical properties of the mineral can be used to understand the conditions of deformation of the rock concerned.

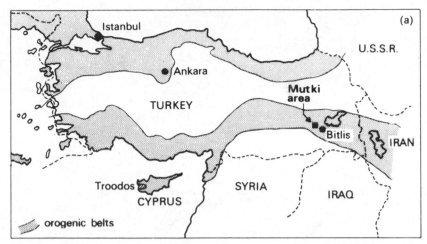

Figure 4.1 Location of Mutki in Turkey.

The Mutki granitic mylonite, Turkey

The example to be described here is from south-east Turkey, near the city of Bitlis (Fig. 4.1). This region is the site of present-day plate collision, between the Arabian Plate, which is moving northwards at an average rate of 40 mm per year relative to the Asian Plate. The collision is thrusting the Arabian Plate beneath the Asian Plate, and the thrust plane concerned moved during an earthquake in 1973 (Sengör & Yilmaz 1981). The over-riding part of the Asian Plate has been up-arched in a series of anticlines, which display older thrust horizons within them. One such thrust appears near the village of Mutki, 27 km west of Bitlis, and has been described by Hall (1976, 1980). The oldest rocks in this area are Devonian to Permian sediments, which underwent regional metamorphism and were intruded by granites in late Palaeozoic times. At Mutki, one such granite intrusion has been thrust above a mixture of fragmented rocks of Mesozoic age (Hall 1980) (Fig. 4.2). The thrust surface dips generally northwards, with granite above and frag- mented Mesozoic rocks below.

As the southern thrust contact of the granite is approached, dynamic metamorphic processes become more marked. This is seen in the field as a change from massive, medium-grained granite to foliated granitic gneiss. The change occurs by the increasingly frequent appearance of shear zones within the originally homogeneous granite, with foliated gneiss occurring in and adjacent to the shear zones. In the inner parts of the shear zones, where the granite is most intensely altered, it has changed into a fine- grained, strongly foliated mylonite (c.f. Berthe et al. 1979). Thus the progressive dynamic metamorphic sequence observed in the field is

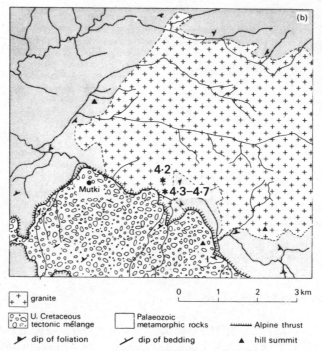

Figure 4.2 Geological map of the Mutki area, with location of specimens shown in Figures 4.3 to 4.7 indicated.

plagioclase

perthite

Figure 4.3 Granite slightly affected by dynamic metamorphism. Scale bar 1 mm.

granite→granitic gneiss→mylonite. The change can be observed over about 10 m adjacent to individual shear zones, and in the granitic intrusion as a whole it becomes apparent over about 1 km as shear zones become more frequent towards the base of the overthrust granitic intrusion.

Thin section study shows that even in granite which appears in the field to be unaffected by dynamic metamorphism, some metamorphism can be seen in thin section. Figure 4.3 shows the most unaltered granite studied. It consists of approximately equal proportions of K-feldspar, plagioclase feldspar (An$_{23}$) and quartz, with a minor proportion of blue-green hornblende. These proportions of minerals make it an adamellite according to the QFS classification scheme for granitic rocks. The K-feldspar is perthite, consisting of host orthoclase containing platey or ribbon-shaped inclusions of untwinned plagioclase. The quartz crystals show strain shadowing of their extinction between crossed polars, the boundaries between quartz crystals often have a lobate form. The hornblende consists of fine-grained aggregates which are apparently **pseudomorphs** after larger primary hornblende crystals. The plagioclase crystals have bent twin lamellae.

Figure 4.4 shows a coarser-grained granite, with little evidence for dynamic metamorphism in the hand specimen similar to the sample in Figure 4.3. The grain boundaries between quartz crystals are again often lobate, but in many places small new crystals of quartz are developed between both quartz and quartz, and quartz and feldspar crystals. This granulation of grain boundaries is a common feature of deformed massive rocks containing

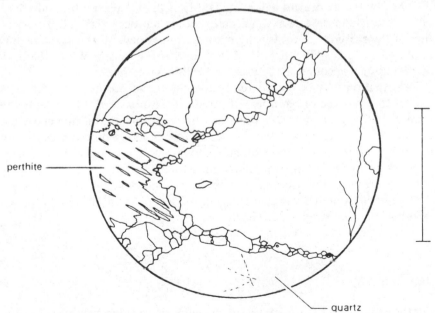

perthite

quartz

Figure 4.4 Granite with mortar texture. Scale bar 1 mm.

quartz and feldspar and is known as **mortar texture**. The small grains are not strained, have planar boundaries with one another (in contrast to the lobate boundaries of the large grains) and develop 120° triple junctions. Thus the original igneous grain boundaries appear to have been fragmented and recrystallized.

The evolution of the lobate grain boundaries and the mortar texture may be understood by considering the effect of the deformation of the original granite. This occurred in Cenozoic times, long after the solidification of the granite during Palaeozoic times. At the relatively high level in the lithosphere at which the Alpine thrusting occurred, the quartz and feldspar crystals were strong because the temperature was low, and the weakest part of the polycrystalline granite was the boundaries between crystals. The parts of the crystals adjacent to these boundaries were more deformed than the interiors, and along the boundaries some parts of the crystals were more strongly deformed than others. As a result, small crystals formed along grain boundaries.

It is worth saying more about the processes of deformation, and the textures which result. We have seen that mineral assemblages of metamorphic rocks may be interpreted using the Phase Rule to obtain a value for the temperature and pressure of metamorphism, assuming that thermodynamic equilibrium was achieved at some time during the evolution of conditions along the P–T–t path. The interpretation of textures to reveal the conditions of metamorphism is more difficult, because more factors are involved. The texture depends not only upon the mineral assemblage and texture which were present before metamorphism, but also on the amount of deformation, and the rate at which deformation has occurred. A full description of the methods of measuring deformation is beyond the scope of this book (see Means 1976, Nicholas & Poirier 1976, Evans & White 1984). A very simplified account of a few aspects will be given here.

Deformation may be considered as the change of position and shape of part of the rock, before to after deformation (Hobbs *et al.* 1976). Only the change of shape of the rock, which is known as the **strain**, influences the texture. Strain is related to the stresses or forces acting on the rock during the deformation episode, but this subject will not be discussed here. Let us consider a simple case of strain in 2 dimensions only (Fig. 4.5).

In this example, the strain may be described as the change in the length of the two lines l and m, from l_0 and m_0 before deformation, to l_1 and m_1 after deformation. The strains in each case are:

$$e_l = (l_1 - l_0)/l_0$$

$$e_m = (m_1 - m_0)/m_0$$

For the example in Figure 4.5, e_l will have a positive value, and e_m a negative value. If there is no strain, e will be zero, but if the rock sample is highly

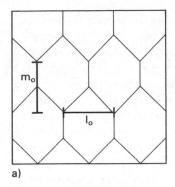

a)

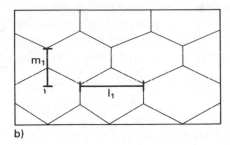

b)

Figure 4.5 Strain of a rock with granoblastic texture (schematically represented by hexagons). (a) Before strain (b) After strain. Further discussion in the text.

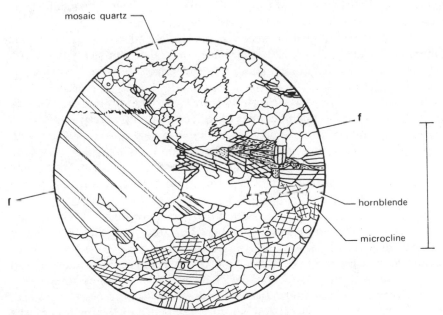

mosaic quartz

f

f

hornblende

microcline

Figure 4.6 Granitic mylonite. f–f foliation plane. Scale bar 1 mm.

strained e will have a large positive or negative value. If the undeformed rock contains several crystals, as shown in Figure 4.5a, they will have to change their shape and perhaps their size during deformation to the shape in Figure 4.5b. In the rocks illustrated in Figures 4.3 and 4.4, the change has been accommodated by the development of small grains along the pre-existing grain boundaries. In these rocks the proportion of new grains, formed during deformation accompanied by dynamic metamorphism, is quite a small proportion of the volume of the rock. However, in the rock illustrated in Figure 4.6 the proportion of new grains has become much larger. The distinction between the new crystals (or other material, as we shall see) and the pre-existing crystals is central to the classification of rocks produced during dynamic metamorphism. Figures 4.3 and 4.4 show deformed (or 'crushed') granite, while Figure 4.6 shows a sample of **mylonite**. It can be seen that some of the original crystals of the granite survive as **porphyroclasts**, while the matrix in between is made up of smaller crystals. Large variations of grain size, as shown in this sample, are characteristic of dynamic metamorphic rocks. The nature and proportion of the matrix between the porphyroclasts is central to the classification of mylonitic rocks. The mylonite in Figure 4.6 comes from within a shear zone. The original crystals of the granite are now separated by an appreciable volume of matrix, which has a distinct foliation f–f. The surviving porphyroclasts of quartz and feldspar have a roughly lenticular shape. The presence of foliation in the matrix, and of lenticular porphyroclasts, are both characteristic features of mylonites.

The textures developed between the grains depend not so much on the amount of strain as on the rate of strain. If rocks deform quickly (i.e. strain rates are large), quite different textures are produced from those which arise if they are deformed slowly. In rocks near major fault and thrust planes, the amount of strain is very variable (Figure 4.7). The strain is said to be heterogeneous. Assuming that parts of the rock with large strains and parts with small strains were deformed during the same time period, the parts with large strains will have experienced higher strain rates than the areas with small strains. But the amount of strain and the rate of strain are not the only controls on the texture, temperature and pressure, (which is related to the deforming stresses), are important too.

In general, two types of deformation mechanism compete. At high strain rates and low temperatures, individual crystals are strong, and they tend to become stronger as deformation progresses, due to processes called **work hardening**. The grains break up mechanically, a process called cataclasis, and smaller grains form at their boundaries, as we saw in Figures 4.3 and 4.4. Thus dynamic metamorphism tends to make the rock become finer grained, on average. By contrast, at low strain rates and high temperatures, the crystals are weaker, and as a result are unstrained and tend to be coarser grained. These strain-rate related differences in the matrix between porphyroclasts are the ones which are used in classification.

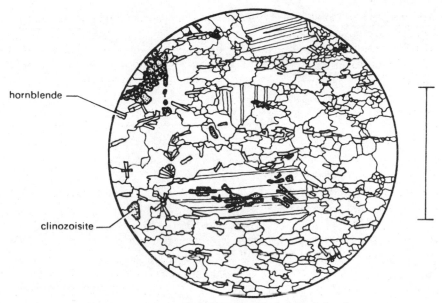

Figure 4.7 Most strongly deformed mylonite, Mutki, Turkey. Scale bar 1 mm.

If strain rates are very high indeed, the rocks actually melt, because work hardening causes large amounts of energy to be stored in the crystals as heat. Rapid cooling may preserve the melt as a glass, which is called **pseudotachylyte** or hyalomylonite. Less rapid cooling will cause the pseudotachylyte to crystallize as very fine-grained, flinty rock, but its original character can often be recognized from field relationships. At slightly lower strain rates, or higher temperatures and lower rates of cooling, the matrix will be crystalline. If the crystals are microscopic (< 0.1 mm), the rocks is called a mylonite, like the example in Figure 4.6. If the crystals are larger, the rock is known as a **blastomylonite**. Table 4.1 presents a classification of mylonites and related rocks, based upon the proportion of matrix, and its character.

In the multi mylonites, the minerals of the matrix differ significantly from the porphyroclasts. Whereas the K-feldspar in the original granite was orthoclase perthite, the K-feldspar crystals in the matrix are of microcline, without inclusions of albite. The plagioclase porphyroclasts can be seen to be strained by their bent twin lamellae. The quartz in some porphyroclasts consists of many crystals adjoining in lobate boundaries. The extinction positions of these crystals differ by angles of up to 5°. They were originally parts of one large quartz crystal, parts of which were rotated relative to one another during deformation. At first they may have had gradational boundaries, giving rise to a large, strain-shadowed crystal. As deformation proceeded, followed by annealing, the differently oriented domains in the original large porphyroclast developed into separate crystals with lobate boundaries. A striking feature of this mylonite is the variety of grain sizes,

Table 4.1   Textural classification of mylonites (Spry 1969, Sibson 1977).

Matrix	Random fabric	Oriented fabric
Incohesive	Fault breccia (clasts < 30%) Fault gouge (clasts > 30%)	
Cohesive, fine grained, 0–10%	Crush breccia	Coarse grain size > 5 mm Fine 0.1–0.5 mm Micro- < 0.1 mm
Cohesive, fine grained, 10–50%	Protocataclasite	Protomylonite
Cohesive, fine grained, 50–90%	Cataclasite	Mylonite
Cohesive, fine grained 90–100%	Ultracataclasite	Ultramylonite
Cohesive, visibly crystalline		Blastomylonite (could be divided into categories by proportion of matrix, like mylonite series)
Glassy, 100%	Pseudotachylyte	Hyalomylonite
Phyllosilicates		Phyllonite (could be divided into categories by proportion of matrix, like mylonite series)

*Empty spaces indicate either that rocks in the category have not been described and probably do not occur, or that they would not be described as dynamic metamorphic rocks.

grain boundary types and grain shapes seen in a single thin section. Such textural heterogeneity is characteristic of dynamic metamorphic rocks.

Figure 4.7 shows an even more strongly deformed mylonite from one of the shear zones in the granite. It is darker in colour than the previous mylonite, and the matrix is obviously foliated. The porphyroclasts are not granite, but fine-grained sandstone, vein quartz and impure limestone, derived from the regionally metamorphosed sediments into which the granite was intruded. The tectonic movements along the shear zone have brought highly deformed country rocks into the granite. This introduction of *exotic* material is a feature of faults and thrusts with large displacements.

The Lochseiten Mylonite, Switzerland

The tectonic evolution of the Alps is unusually well understood. In the example to be described here, mylonite formation has played a major role in the mechanism of thrusting. A detailed account has been given by Schmid (1975).

The Lochseiten Mylonite is a layer of marble 1–2 m thick which lies at the base of the Glarus Nappe of the Alps to the southeast of Zurich, Switzerland. It is in the external zone of the Alpine fold belt, where the Mesozoic sedimentary rocks are either not metamorphosed, or display very low-grade regional metamorphism (Chapter 5, Frey 1987). It is a classic area for tectonic research, in which Escher von der Linthe, Bertrand and Heim first established the existence of large-scale overthrusting in any orogenic belt. Readers who wish to know more are referred to Bailey (1935).

The Lochseiten Mylonite lies between the lowest nappe of this part of the Alpine chain, the Glarus Nappe, and untransported basement rocks, the **autochthon**. The autochthon is the upper part of the European Plate, deformed but not transported during the Alpine earth movements. At the time of thrusting the nappe was a cold, comparatively rigid slab of rock about 2.5 km thick, mostly made up of a Permian conglomerate. The autochthon was also rigid, formed of early Cenozoic rocks which had previously been altered to slate by regional metamorphism (Chapter 5). The mylonite is a thin layer between the two which apparently acted as the

Figure 4.8   Lochseiten Mylonite, Schwanden, Canton Glarus, Switzerland. Scale bar 1 mm.

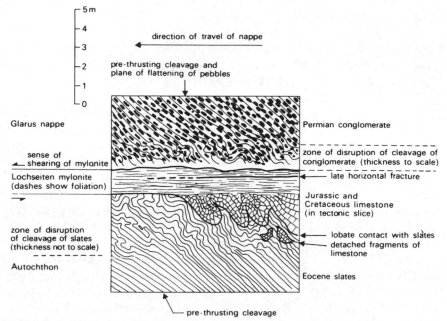

Figure 4.9 Generalized cross section through the Lochseiten Mylonite zone at the base of the Glarus Nappe, from Schmid (1975).

lubricant of the thrust plane, allowing the Glarus Nappe to travel at least 35 km over the autochthon along the thrust.

Figure 4.8 shows a thin section of the mylonite, at a low magnification. It has a well-marked but uneven foliation, which can be seen in the field to be sub-parallel to the plane of the thrust. It has a typical mylonitic texture of porphyroclasts of marble set in a matrix of calcite crystals, which are too fine grained to be distinguished in the drawing. The calcite crystals of the porphyroclasts are much larger, but are strained. Many of the porphyroclasts are recognisable fragments of calcite veins, cross-cutting the fragmental texture of earlier formed mylonite. These veins were themselves later fragmented, so that some specimens of mylonite consist entirely of more or less fragmented veins of calcite in the fine-grained matrix. The matrix calcite crystals are not strain shadowed, have flat grain boundaries and 120° triple junctions.

Figure 4.9 shows a generalized vertical cross section through the mylonite zone. From a study of the minor structures near the thrust zone, Schmid concluded that the majority of the relative movement between the nappe and the autochthon took place in the metre-thick layer of mylonite, the deformation of the basal part of the conglomerate above the thrust and the slate below, and the fault displacement on the late flat-lying fracture only accounting for a small fraction of the 35 km displacement. Most of the movement was taken up by laminar flow in the mylonite layer, during which

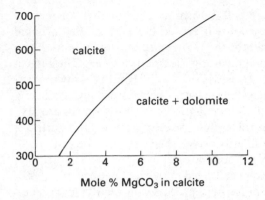

Figure 4.10 Phase diagram showing the univariant curve for the composition of calcite co-existing with dolomite at 0.2 GPa pressure. If the calcite contains 6% $MgCO_3$, it crystallized at 560°C.

the foliation developed. The mylonite was derived from small thrust slices of Jurassic and Cretaceous limestone, which were over-ridden by the thrust (a part of such a slice is shown in Fig. 4.9). The marble spread out like butter over bread, along the thrust plane, because it was very much more plastic than either the slate or the conglomerate at the time of thrusting. The presence of the broken calcite veins shows that the plastic deformation of the mylonite was assisted by solution of calcite in intergranular fluid and its subsequent re-precipitation in veins. The granoblastic matrix indicates that recrystallization, rather than mechanical deformation of grains was the dominant flow mechanism.

The small crystals of calcite in the matrix of the mylonite contain a proportion of magnesium, and a small proportion of dolomite ($CaMg(CO_3)_2$) crystallized as well as calcite ($CaCO_3$). In such a case, the temperature of crystallization of the two minerals can be calculated, because as the temperature rises the proportion of magnesium in calcite and of calcium in dolomite, also increase. The relative proportions of the two minerals are shown in a phase diagram in Figure 4.10. There is a **solvus** between calcite and dolomite. Provided that the magnesian calcite and dolomite crystallized in equilibrium, which appears likely for the matrix of the Lochseiten Mylonite, the temperature of crystallization can be determined. The exact method used in this type of **geothermometer** has been simplified here, because the exact positions of the univariant curves on the phase diagram of Figure 4.10 depend on pressure and fluid composition, but the principle is as explained here.

Dynamic metamorphism in relation to depth

In the two examples we have considered, it is seen that the nature of the matrix, used to classify mylonitic rocks, has been influenced by the temperature at the time of metamorphism. As discussed in Chapter 2, temperature in turn depends upon depth of burial in the crust. Rocks at shallow levels will tend to be cold and strong. They will deform in fault or thrust zones by fracture, and by mechanical breakdown (i.e. cataclasis) of the original crystals of the rock. At greater depths the matrix will crystallize, very likely aided by the presence of fluid escaping along the fracture zones. The subject has been reviewed by Sibson (1977), who draws a contrast between the mechanical behaviour of rocks near the surface, which deform by elastic and frictional processes, and those at greater depths, which deform in a quasi-plastic fashion. The near surface rocks thus develop narrow fault surfaces, while the deeper rocks develop slightly wider shear zones. The pattern of deformation with increasing depth is summarized in Figure 4.11.

This diagram is descriptive, and generalized. The exact pattern of types of deformation mechanism and metamorphic processes in the dynamic meta-morphic rocks will vary according to the intrinsic mechanical properties of the rocks (i.e. it will be different in granites and in shales), the geothermal gradient, the rate of deformation, whether or not fluid is present in the fractures, and so on. With these provisos, it provides a very valuable conceptual model for understanding dynamic metamorphism associated with fractures in the Earth's crust. The principal feature which it shows, the transition from elastico-frictional deformation processes near the surface, to quasi-plastic ones (associated with metamorphism) at greater depth, has been demonstrated by numerous field studies.

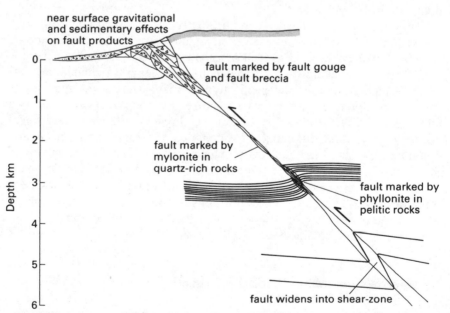

Figure 4.11 Variation of the nature of dynamic metamorphism with depth, on a thrust fault. The actual depths at which the different types of rock form vary greatly according to the nature of the rocks, the geothermal gradient, and the rate of strain on the fault.

5 Metamorphism associated with lithospheric extension

Metamorphism at spreading oceanic ridges

The Earth's oceanic crust has formed by extension from oceanic ridges. In oceans where subduction zones are not yet extensively developed, such as the Atlantic, these ridges are close to the centre of the ocean basin. In oceans such as the Pacific, which are surrounded by active subduction zones, the ridges are not central. Oceanic crust and the upper mantle, together constituting the oceanic lithosphere, are constructed by extension at the ridges. Geophysical studies indicate that oceanic ridge extension creates new crust symmetrically on either side of relatively narrow extensional rift regions, often marked by topographical graben structures. The oceanic crust so formed is mostly composed of basic igneous rocks (basalts when erupted on the ocean floor, dolerites and gabbros when intruded into previously constructed oceanic crust), and the oceanic mantle of ultrabasic rocks.

Dredging of rocks from oceanic fracture zones, and some drilling undertaken in the international ocean-floor drilling project, has revealed that metamorphic rocks, principally **greenstones** and more rarely amphibolites, are frequently present, formed by metamorphic processes associated with the construction of ocean floor itself. However, field relationships between these metamorphic rocks and their unmetamorphosed equivalents are not clear, because of the limited sampling which is possible, and the metamorphism of ocean-floor rocks can be made clearer by studying ancient fragments of ocean floor which are now found on land.

Many orogenic belts preserve remnants of oceanic crust and mantle, which have not been pushed down into the mantle at subduction zones, but instead have been emplaced by **obduction** onto continental crust. These remnants are called **ophiolites** (Gass & Smewing 1981, Mason 1985). Some ophiolites preserve an undisrupted sequence downwards from deep-water sediments deposited on the ocean floor, through the oceanic crust into the upper mantle. One of the most famous of these ophiolite sequences is found in the southeastern part of Cyprus, in the Troodos Mountains (Fig. 5.1). The Troodos Ophiolite is one of a number of well-preserved fragments of oceanic crust and mantle occurring near the southern flank of the Alpine–Himalayan orogenic belt between the Aegean Sea and the Straits of Hormuz.

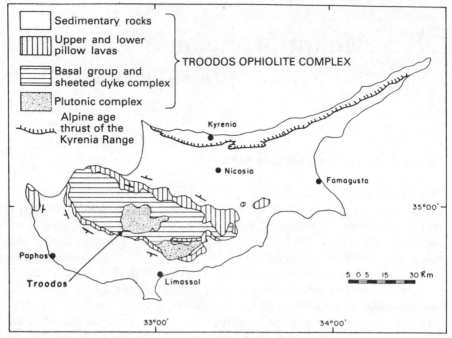

Figure 5.1 Geological map of Cyprus, showing the Troodos Ophiolite.

The Troodos Ophiolite Complex, Cyprus

The Troodos Mountains have an anticlinal structure, with the ophiolite complex occurring in the core of an E–W trending anticline (Fig. 5.1). The overlying sediments show that the ophiolite was deeply covered by sea water immediately after the oceanic crust was formed, in late Cretaceous times, and this oceanic basin was filled by calcareous turbidites in latest Cretaceous and early Cenozoic times. Like all ophiolite complexes, the Trodos Ophiolite has a distinctive layered structure, and the number, thickness and geophysical properties of its layers match those of oceanic crust closely (Fig. 5.2). The ophiolite and its covering sediments have been overthrust by sediments from the Turkish continental margin, which form the Kyrenia range of mountains of northern Cyprus. There is probably also another major thrust beneath the ophiolite, underlain by continental crust, but this structure does not come to the surface, although there are some minor thrusts within the ophiolite. The updoming of the anticline took place in Pliocene–Pleistocene times, forming the present mountainous area. At least part of the updoming is due to hydration of the ultrabasic rocks of the ophiolite, to serpentine. This process caused a large decrease in density, from about 3300 kg m^{-3} to 2550 kg m^{-3}. The serpentinite rock formed by hydration is relatively weak, compared with the parent ultrabasic rocks, and has therefore moved upwards due to gravitational forces to form a diapiric plug, similar to a salt dome in a sedimentary basin. This plug has been breached by erosion near the top of the Troodos Mountains, and veins of fibrous serpentine in it have been quarried as asbestos.

There are two types of metamorphic rock present, metamorphosed crustal rocks, which were originally basic in composition, and metamorphosed mantle rocks, which are ultrabasic in composition. The rocks shown in the column section (Fig. 5.2) with its distinctive ocean-floor sequence, are variably affected by metamorphism. The basaltic rocks highest in the sequence, which are petrographically and geochemically distinct from those below, are known as the Upper Pillow Lavas. They show a limited amount of low-grade metamorphism, which will be described in more detail later on. In the field, they are dark in colour, contrasting with the more metamorphosed Lower Pillow Lavas, which have changed colour from black to green (often pale green). These rocks are true greenstones, retaining very impressive pillow structures, and showing no sign of foliation. Most of the underlying Sheeted Complex of multiple dykes is also metamorphosed to greenstone, although occasionally darker green amphibolites occur near the base of the complex. Metamorphism becomes much less widespread in the gabbro layer.

Figure 5.3 shows a thin section of a metamorphosed basalt from the Upper Pillow Lavas. It contains primary lath-shaped crystals of plagioclase feldspar, which survive from the original crystallization of the igneous magma, and are unmetamorphosed. It also contains phenocrysts of olivine,

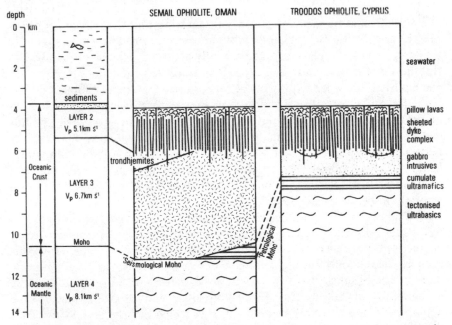

Figure 5.2 Comparative column sections of the structure of oceanic crust and upper mantle, determined by seismic methods, the Semail Ophiolite, Oman, UAR, and the Troodos Ophiolite.

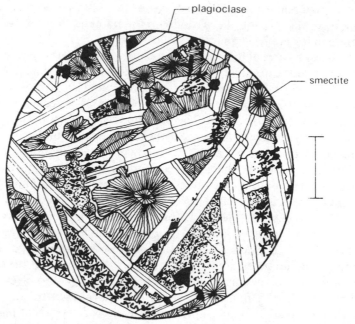

plagioclase

smectite

Figure 5.3 Slightly metamorphosed basalt, from the Zeolite Facies zone, Skouriotissa, Troodos Ophiolite. Scale bar 0.1 mm.

again unaffected by metamorphism, but showing that the primary igneous magma was picritic rather than basaltic in composition. There was originally a matrix of basaltic glass between the olivine and plagioclase crystals, but this has devitrified to radiating crystals of **smectite** which are intergrown in some parts of the rock with dusty grains of opaque minerals. 'Smectite' is a name for clay minerals of the montmorillonite group, which is used here in an informal sense for the material replacing the basaltic glass. The crystals can be resolved under high magnifications of the petrological microscope, and are distinguishable from white mica by their pale yellow colour and lower refractive indices. The opaque mineral cannot be identified under the transmitted light microscope, but has been determined by X-ray diffraction as titanium-bearing magnetite.

The sample comes from the massive part of a lava flow, which does not have cracks or inter-pillow spaces containing the zeolite minerals which are characteristic of this part of the Troodos metamorphic sequence. However, the presence of smectite replacing glass is also typical of the lowest-grade part of the sequence. It hardly needs to be said that the textures of these rocks show that they are a long way from either textural or thermodynamic equilibrium, and therefore assemblage lists which may be related to metamorphic conditions cannot be compiled.

With increasing metamorphic grade and depth beneath the ancient ocean floor, smectite is replaced by quartz and chlorite, and the plagioclase feldspars become clouded owing to incipient breakdown. They still have labradorite composition. Epidote and actinolite appear shortly after quartz and chlorite. Figure 5.4 shows a rock from this level in the Lower Pillow Lavas. The mineral assemblage is plagioclase + actinolite + chlorite + quartz + opaques (titaniferous magnetite). This is similar to low-grade basic igneous rocks from collision orogenic belts (Chapter 6), except that the plagioclase feldspar is of different composition. The rock has a texture which suggests that it contains an equilibrium mineral assemblage. Actinolite is a calcic amphibole $Ca_2(Fe^{+2}, Mg)_5[Si_8O_{22}](OH)_2$ with a close resemblance to hornblende. It differs chemically from hornblende in having little aluminium in it. Although it is green and pleochroic under the microscope, similar to hornblende, it can be distinguished by the extinction angle γ: z, which is up to 18°, compared with hornblende's angle of 25–30°. In metamorphosed basic igneous rocks, hornblende takes the place of actinolite as the metamorphic grade increases, and so it is important to distinguish the two minerals, and this can usually be done with the petrological microscope, as in this case. The minerals distinguish metamorphic conditions of the **amphibolite facies** from those of the **greenschist facies**.

With a further increase in metamorphic grade, the actinolite is replaced by

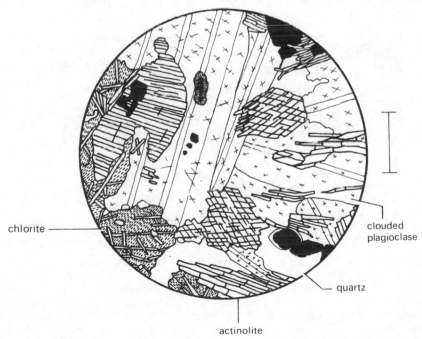

Figure 5.4 Greenstone from dyke in Sheeted Complex, Phterykoudhi, Troodos Ophiolite. Scale bar 0.1 mm.

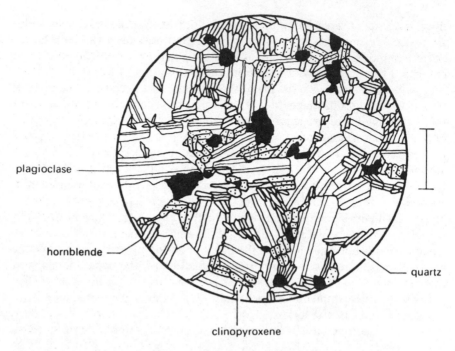

plagioclase

hornblende

quartz

clinopyroxene

Figure 5.5 Amphibolite from dyke in Sheeted Complex, Lemithou, Troodos Ophiolite. Scale bar 0.1 mm.

Figure 5.6 Dykes in the Sheeted Complex at Zoopiye, Troodos Ophiolite, looking north-wards. The dykes are intruded into light-coloured plagiogranite, in the lower part of the Sheeted Complex. The gently dipping dyke is metamorphosed to greenstone, while the near vertical dykes are unmetamorphosed dolerite. Models for the evolution of sheeted complexes by repeated dyke intrusion suggest that initially vertical dykes rotate to a gentler dip as they spread away from the ridge. Thus this drawing suggests that the ocean-floor metamorphism occurred close to the ridge, but not exactly on it.

hornblende and chlorite disappears. Figure 5.5 shows a rock of this grade, which comes from the lower part of the Sheeted Complex. It has the mineral assemblage plagioclase + quartz + hornblende + diopside + opaques. Below this level, the dolerites of the Sheeted Complex are replaced by plagiogranites and gabbros, which are metamorphosed much less frequently. In some localities, field evidence shows clearly that the metamorphism of the dykes in the Sheeted Complex occurred while extension was still going on. Figure 5.6 is a field sketch of a locality near the village of Zoopiye, on the southern limb of the Troodos Anticline, in the lower part of the Sheeted Complex. Dykes have been intruded into slightly older plagiogranite. It can be seen that the earlier intruded dykes are sloping, whereas the younger dykes are nearer to vertical. This can be related to the process of construction of the Sheeted Complex, which tilted as it spread from the oceanic ridge. The older dykes have been metamorphosed to greenstone, while the youngest ones are unmetamorphosed dolerite, showing characteristic spheroidal weathering. This shows that the older dykes had undergone metamorphism, and been tilted from the vertical, before the youngest dykes were intruded.

Below the Sheeted Complex, the igneous rocks of the crustal part of the Troodos Ophiolite Complex are usually not metamorphosed. Figure 5.7 shows a rare exception. This rock comes from a xenolith near the roof of one of the gabbro intrusions in the upper part of the Plutonic Complex. It is coarser grained than the rocks shown in Figures 5.3 to 5.5. It is a typical pyroxene hornfels composed of clinopyroxene and calcic plagioclase feldspar. It shows a high degree of textural equilibrium. There is little doubt about the immediate process which has caused the metamorphism of this rock, it is contact metamorphism due to the heat of the surrounding gabbro intrusion.

Although exact estimates of the temperatures of metamorphism cannot be made for the assemblages with plagioclase feldspar and amphibole which are found in the Troodos Ophiolite, rough values can be obtained, and they indicate a metamorphic field gradient of $> 300°C$ km^{-1}, which is unusually high. Although the Troodos Ophiolite is unusually thin, and appreciably thinner than typical oceanic crust, all ophiolite sequences which were metamorphosed during extension show comparable high metamorphic field gradients. Metamorphic rocks dredged and drilled from ocean floors show similar high gradients. A. Miyashiro pointed out in 1961 that these high metamorphic gradients were matched by high near-surface geothermal gradients measured near oceanic ridges, and that this indicated a close link between ridge-forming processes and ocean-floor metamorphism. Since then, oceanographic, petrological and geochemical research have led to an impressive level of understanding of the process of heat transfer beneath oceanic ridges, and this illuminates the understanding of metamorphic processes generally (Anderson & Skilbeck 1981).

Despite the continuing debate about the proportion of the oceanic crust which has undergone metamorphism, the fact that oceanic crust covers the majority of the outer layer of the Earth's lithosphere means that **ocean ridge metamorphism** is arguably the most important regional metamorphic process on Earth (Elthon 1981).

Metamorphism also occurs at deeper levels in the oceanic crust of the Troodos Ophiolite. The ultrabasic layers in the ophiolite sequence are of two kinds. There are cumulate ultrabasics, which represent the deeper parts of intrusive bodies like the gabbro in which the rock of Figure 5.7 occurs, and beneath these are peridotites of the harzburgite variety (containing orthopyroxene crystals). The latter show a large-scale foliation fabric defined by preferred orientation of olivine and orthopyroxene crystals, and also by extension of chrome-rich spinel crystals. The foliation is gently dipping, parallel to the major boundaries in the ophiolite sequence itself, for example the seismic and petrological 'Mohos'. It formed in the uppermost part of the mantle during the intense deformation of the rocks turning over at high temperatures beneath the spreading oceanic ridge (Fig. 5.8). At this time the upper mantle had heated to temperatures at which basaltic fluids separated from the mantle rocks by partial melting, and geochemical evidence indicates that the foliated harzburgites are the residue left after a significant proportion of the parent ultrabasic rock had been removed by partial melting.

Thus there are two very different types of metamorphism associated with the spreading of the oceanic ridge which formed the Troodos Ophiolite in

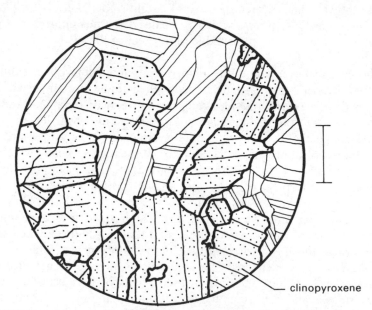

clinopyroxene

Figure 5.7 Pyroxene hornfels from xenolith of basalt in a gabbro intrusion just below the Sheeted Dyke Complex of the Troodos Ophiolite at Ayios Demetrios. Scale bar 0.1 mm.

late Cretaceous times. Metamorphism occurred in the higher parts of the oceanic crust, leading to the formation of hydrous minerals such as smectite, chlorite, epidote and amphiboles and was associated with very high metamorphic temperature gradients. At approximately the same time, at the top of the upper mantle, the solid rocks underwent deformation and metamorphism giving rise to foliation at high temperatures.

Metamorphism did not cease after ridge spreading. There is dynamic metamorphism associated with movements which occurred on transform faults around the time of spreading, and later dynamic metamorphism

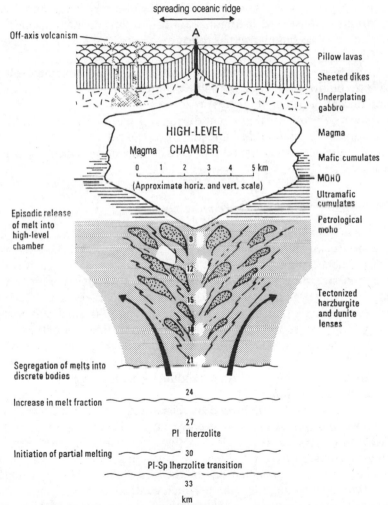

Figure 5.8 Model for the structure and evolution of a spreading oceanic ridge, from Gass & Smewing (1981). The shape of the magma chamber is very speculative. The nature of the different layers which go to form an ophiolite is shown on the right.

associated with the process of upthrusting, known as obduction, which brought the ophiolite above continental crust. Finally, there has been the hydration of peridotite to serpentinite, which has been partly responsible for the uplift of the Troodos Mountains. Geological evidence shows that obduction occurred in latest Cretaceous times, very shortly after the oceanic crust had formed by spreading from the oceanic ridge, whereas serpentinization appears to have occurred more recently. The precise tectonic history of the Troodos Ophiolite is still being worked out, but it is now generally agreed that it probably formed above a subduction zone, rather than at a mid-oceanic spreading ridge.

The metamorphism of the higher levels in the oceanic crust, oceanic ridge metamorphism, illustrated in Figures 5.3 to 5.6, has distinctive characteristics which may be summarized as follows:

1　There is a progressive metamorphic sequence from unmetamorphosed basaltic pillow lavas to amphibolites, which developed during the construction of the ocean floor by spreading.
2　The average geothermal gradient during this period was high, similar to the gradients at present-day spreading oceanic ridges.
3　Most of this metamorphism did not lead to the development of tectonic fabrics.

Hydrothermal activity near spreading ridges

Studies near spreading oceanic ridges have shown that sea water circulates through hot rocks, and emerges at submarine brine springs. Where the water is hottest, metal sulphides are precipitated near the spring to form the chimney-like structures called **black smokers**. The hot water supports unique colonies of living organisms around the springs. The intervals at which hot springs occur (~ 3 km apart) indicate that they are the emergent parts of relatively large convection cells of hot water, circulating through the rocks adjacent to the ridges. These discoveries confirm the conclusions of ore mineral investigations, and geochemical studies, which also show that metamorphism and ore genesis near spreading oceanic ridges occur due to hydrothermal circulation.

This conclusion is also supported by measurements of heat flow through ocean floor sediments. Away from oceanic ridges, these measurements show that heat flow becomes steadily less as the age of oceanic crust increases, showing that much of the oceanic crust is cooling by conduction of heat through the oceanic lithosphere up to the ocean floor. This is a marked contrast with continental lithosphere. Much continental lithosphere is thousands of millions of years old, and the geothermal gradients in it have arrived at a steady state. All oceanic lithosphere, on the other hand, is less than 80 million years old, and is still cooling from the relatively hot condition, with

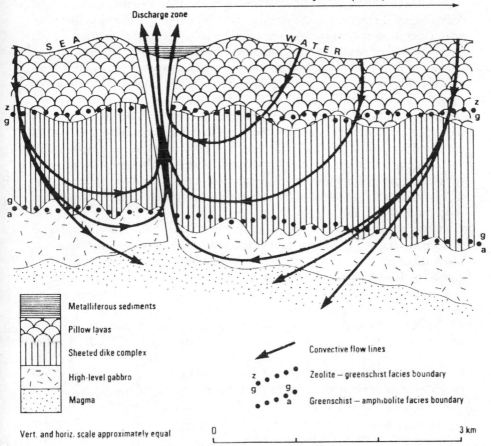

Radius of recharge flow   (ca.3 km)

Discharge zone

SEA   WATER

Metalliferous sediments

Pillow lavas

Sheeted dike complex

High-level gabbro

Magma

Convective flow lines

z • • • •   Zeolite – greenschist facies boundary
g   g

g • • •   Greenschist – amphibolite facies boundary
•   a

Vert. and horiz. scale approximately equal

0   3 km

Figure 5.9 Hydrothermal circulation pattern of sea water near an oceanic ridge, from Gass & Smewing (1981). The boundaries of the metamorphic facies formed by the alteration of the rocks by the sea water are shown.

high geothermal gradients, which occurs at oceanic ridges. Heat-flow measurements made near oceanic ridges show very large variations, from some of the highest values measured anywhere near the Earth's surface to low values. These variable values are presumably related to the existence of submarine water circulation systems. Where cool sea water is being drawn down into the convection cells (Fig. 5.9), the near-surface heat-flow value is low, where there is an upwelling hot current, close to one of the submarine brine springs, the near-surface heat-flow value is high. The pattern of oceanic ridge metamorphism, and the hydrogen and oxygen isotopic compositions of the greenstones and amphibolites, show that the hydrothermal circulation systems extended down through the oceanic crust as far as the top of the gabbro layer.

The form of the hydrothermal circulation cells is important for the temperature distribution near oceanic ridges. The parts of the cells where cool water is permeating downwards cover a large area, whereas the uprising parts of the cells cover a smaller area (and therefore the flow rate is correspondingly higher). The uprising brine columns often follow fault systems. This is indicated by the relationship of black smokers to fault scarps on present-day oceanic ridges, and also by the association of copper ore deposits with ancient graben fault systems in Cyprus. This pattern means that the metamorphic boundaries, between low-temperature zeolite-bearing lavas, middle temperature greenstones and high temperature amphibolites are irregular. This can be seen in Cyprus, where the boundaries between rocks metamorphosed in zeolite, greenschist and amphibolite metamorphic facies lie at different levels in the ocean-floor sequence in different parts of the Troodos Ophiolite.

In the upper parts of oceanic lithosphere, where oceanic ridge metamorphism occurs, extension takes place by **non-penetrative** mechanisms, extensional graben-type faulting at the shallow level (accompanied by surface extrusion of lavas), and multiple dyke intrusion at a deeper level. Between the faults and dykes, the rocks were not undergoing internal deformation during metamorphism, and this accounts for the lack of tectonite fabrics.

The overall model for spreading and heat transfer at oceanic ridges is presented in Figure 5.10. It accounts well for the observed facts on metamorphism in ophiolite complexes, rocks dredged and drilled from the ocean floors, and measurements of heat flow near oceanic ridges. It makes it perfectly clear that regional metamorphism is not confined to regions of

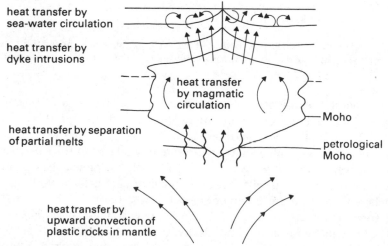

Figure 5.10   Different mechanisms of heat transfer beneath an oceanic ridge. This diagram is to the same scale as Figure 5.8. Note that conduction is not a process which transfers heat over distances of more than a few metres.

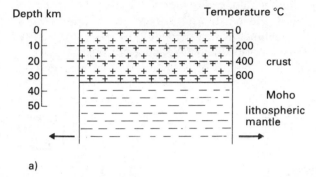

Depth km   Temperature °C

crust

Moho
lithospheric
mantle

a)

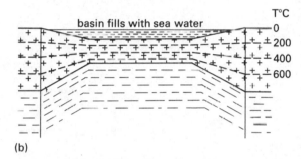

basin fills with sea water

T°C

(b)

Figure 5.11 Evolution of a basin, filled with sea water, by extension of continental lithosphere. (a) Before extension. (b) After extension. Isotherms are only shown in the crust. Extension brings the isotherms closer together, thus increasing the geothermal gradient.

plate convergence or subduction. On the contrary, this most widespread type of metamorphism on Earth is intimately associated with the most extensive plate construction.

Metamorphism in sedimentary basins

Extension of continental areas leads to thinning of the continental crust. Where the extension has been extensive, oceanic crust may form, but in other areas the continental crust remains. Due to isostatic readjustment, the continental crust tends to subside during or after it has been stretched, and therefore becomes a sedimentary basin (McKenzie 1978). Figure 5.11 shows the sequence of events when a prism of lithosphere with continental crust is extended in this way. Geological evidence from actual extensional basins, such as the North Sea, shows that the extension and flooding were relatively rapid processes. The isotherms within the lithosphere would have acted as passive 'marker surfaces' during the extension, so as the crust and upper mantle were stretched and thinned, they moved closer together. This resulted in a perturbed geotherm, with a higher gradient than the steady-state continental geotherm. This prediction has been confirmed by measurement in extensional basins.

The high geothermal gradient would lead to high values for heat flow, and therefore the sediments filling the basin would undergo heating. Where basins are filled by thick sediments, this heating would produce sufficiently high temperatures to cause metamorphism. The sequence of events in such a sedimentary basin, located above a slab of continental crust extended as

Figure 5.12 Sequence of events in the sedimentary basin formed by the filling of the basin in Figure 5.11. (Letters are in the same sequence as in Figure 5.11). (a) Steady-state continental geotherm before extension. (c) Basin has filled quickly with cold sediment after extension. (d) High geothermal gradient migrates upwards and heats sediments in basin. (e) Geothermal gradient relaxes back to steady-state. Asterisk – position of sedimentary rock whose d-T-t path is shown in Fig. 5.13.

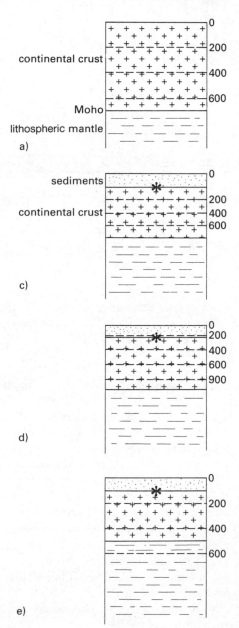

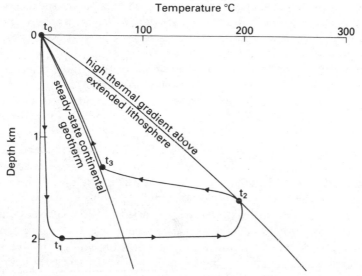

Figure 5.13 Depth–temperature–time path for sediment at the bottom of the basin (letters as in Figures 5.11 and 5.12). t_0 sediment deposited (b), t_1 sediment buried as basin fills quickly (c), t_2 maximum temperature reached as perturbed thermal gradient migrates upwards into basin (d), t_3 temperature falls as thermal gradient relaxes to steady-state value (e).

shown in Figure 5.11, is shown in Figure 5.12. Depth–temperature curves are shown in Figure 5.13.

Along the northern and western flanks of the Alpine mountain belt of Europe, Mesozoic and Cenozoic sedimentary rocks lie above rocks which were metamorphosed during the earlier Caledonian and Variscan orogenic episodes. The complex mountain-building processes buried some of these rocks deeply, and deformed them, so that they have become regional metamorphic rocks. Rocks which were less deeply buried remain essentially sedimentary in their character. The uplift and erosion which occurred during the later parts of the Alpine orogenic sequence have exposed the resultant sequences from sedimentary to regional metamorphic rocks for study (Frey 1987). Because there are several kilometres of vertical relief, it is possible to gain an idea of the distribution of metamorphic rocks in three dimensions, providing a number of valuable case histories to compare with theories of metamorphism associated with crustal convergence.

Sedimentary geologists recognize a number of changes which alter the primary products of deposition, on the bed of the sea or on land, into more or less consolidated sedimentary rocks. These processes are collectively known as **diagenesis**. They may occur very soon after the deposition of the sediment, before it has undergone much burial (i.e. *penecontemporaneously*) or they may occur later, at greater depths of burial. Typical diagenetic changes are the alteration of unconsolidated calcite mud to consolidated limestone, or of unconsolidated turbidite sediments to consolidated grey-wackes. There is no hard-and-fast distinction to be drawn between deep diagenesis and shallow regional metamorphism, the processes concerned merging gradually into one another.

In the northern and western flanking regions of the Alpine orogenic belt, the older metamorphic rocks and the Mesozoic and Cenozoic sediments overlying them have been transported westwards and northwards on a number of gently-dipping thrust planes, giving rise to piles of thrust sheets. Each individual thrust-sheet is known as a nappe. The lower nappes of the Alpine pile are collectively grouped into a major tectonic unit, the Helvetic nappes, or Helvetides. Two rock types predominate among the sediments: limestones and shales. Metamorphism causes little change to the mineral assemblages of the carbonate rocks until high metamorphic grade, but the minerals in the shaley sediments are very sensitive to increases in tempera-ture in the low range associated with deep diagenesis and metamorphism. These pelitic rocks are therefore the ones which are studied by both sedi-mentary and metamorphic geologists to try to understand the transition from diagenesis to metamorphism. They range from unconsolidated shales, through slates to phyllites.

The minerals which change most during metamorphism in their texture, crystal structure and chemical composition are the phyllosilicates. The changes in these minerals can only be accurately determined in the laboratory

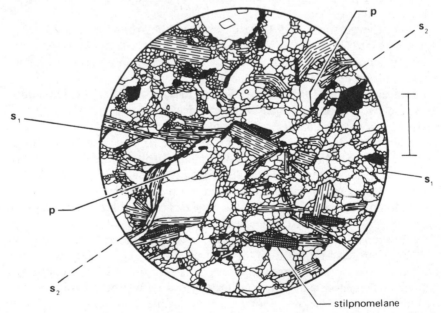

Figure 5.14 Thin section of roofing slate from Sernftal, Canton Glarus, Switzerland. Scale bar 0.1 mm.

by X-ray diffraction and electron microscopy, but the gross changes are clearly visible in hand specimen, and study of thin sections under the petrological microscope permits many of the minerals to be recognized, and also some conclusions to be drawn about the textural development of the rocks.

Figure 5.14 is a drawing of a thin section of a specimen of roofing slate from Eocene turbidites on the northern flank of the Aar Massif of NE Switzerland. The original sediment was a sandy mud and the bedding is picked out by layers with a higher proportion of quartz grains. Many of these still display their original clastic shapes. The white mica flakes are very small and indistinguishable from muscovite under the petrological microscope. However, X-ray studies show that their crystal structure is significantly different from muscovite, being nearer to the clay mineral illite. The informal name **white mica** will be used for such intermediate layered silicates. The mineral assemblage in this rock is quartz + white mica + calcite + stilpnomelane + opaque minerals. It is therefore a calcareous slate, resembling some of the Furulund Group schists of Sulitjelma (Chapter 7) in its overall composition. Stilpnomelane is a mineral which has not been discussed previously. It is a phyllosilicate of the brittle mica group with a complex chemical formula, rich in iron. It is characteristic of low-grade regional metamorphic rocks, passing into the low to medium grades of **Barrow-type** regional metamorphism (stilpnomelane occurs in the chlorite zone of the Grampian Highlands of Scotland). In the pelitic rocks of the Alps of eastern Switzerland, the incoming of stilpnomelane defines an isograd, which is shown in Figure 5.15. Under the petrological microscope, stilpnomelane is difficult to distinguish from biotite. It is pleochroic from brown to yellow and the flakes have straight extinction with the slow vibration direction parallel to the cleavage. In the rock shown in Figure 5.14 the stilpnomelane can be distinguished because it has a slightly less perfect cleavage than biotite, and does not show the characteristic mottled appearance of biotite, when close to the extinction position between crossed polars. Stiplnomelane does not co-exist with biotite in progressive regional metamorphic sequences in pelitic rocks, but disappears with increasing grade before biotite begins to appear.

There are two distinct grain sizes in the white mica flakes of this rock. The larger grains define a penetrative slaty cleavage which cross-cuts the bedding, although this cannot be seen in Figure 5.14. The smaller grains have a less strong preferred orientation. There is a later, strongly defined **spaced cleavage** in the rock, and this is parallel to the axial surfaces of a set of minor folds. In the outcrop, these folds can be seen to fold both the bedding and the earlier slaty cleavage. Some clastic quartz grains show **pressure solution** (Borradaile *et al.* 1982) on the late cleavage surfaces (points **p** in Fig. 5.14). Although in general the quartz grains have not recrystallized during metamorphism, it is apparent from the shapes of grains adjacent to

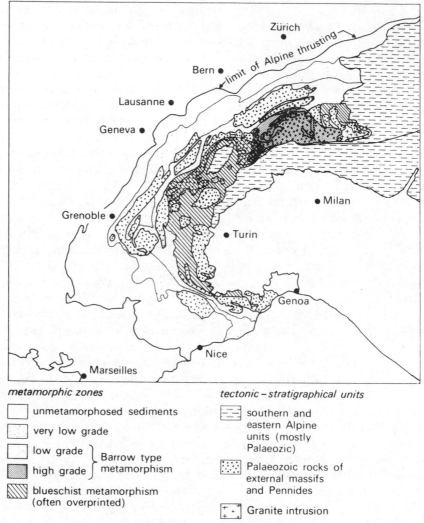

Figure 5.15 Map of metamorphism in western and central Alps. The outer limit of the very low-grade zone is marked by the incoming of stilpnomelane (see Figure 5.19).

the non-penetrative second cleavage surfaces that parts of some grains have dissolved away against the surface. The higher stress developed on the grains against the cleavage surfaces promoted the solution of the quartz in the intergranular fluid, which may have been escaping along the surfaces. This mechanism is an important one in the deformation of sediments and in low-grade metamorphism, especially during the prograde part of the P–T–t cycle, although in this particular rock pressure solution continued after the metamorphic peak.

At higher metamorphic grades, the mineral assemblages resemble those

seen in the low-grade parts of metamorphic sequences such as Barrow's
Zones and the Sulitjelma region of Scandinavia. Figure 5.16 shows a rock
of higher metamorphic grade, which occurs immediately above the 'basement' of rocks metamorphosed during the earlier Variscan orogeny. The
mineral assemblage is quartz + plagioclase(An$_0$) + potash feldspar + white
mica + chlorite + graphite. This list is based upon X-ray determinations of
the minerals by C. Taylor (personal communication). The plagioclase is
untwinned, and therefore cannot easily be distinguished from quartz under
the petrological microscope. The rock is different from the one in Figure
5.14 because the clastic sand component contains feldspar as well as quartz,
and the fine-grained groundmass does not contain carbonate. As in the rock
in Figure 5.14, there are two cleavages, the earlier one penetrative, the later
one non-penetrative. Pressure solution of clastic quartz and plagioclase
grains is visible on the later non-penetrative s$_2$ cleavage surfaces. The opaque
graphite has a distinctly tabular habit, with the plates tending to be parallel
to the white mica flakes. The concentration of graphite in the s$_2$ cleavage
surfaces is a pressure-solution effect. Graphite was much less soluble in the
metamorphic fluid than quartz, feldspar and mica, and so was left behind
when those minerals were removed by solution along the s$_2$ cleavage surfaces. The process is analogous to the formation of stylolites in limestones.

It was emphasized earlier that the white mica in these low grade regional
metamorphic rocks is not muscovite. It is a sheet silicate (phyllosilicate) with
some layers having the same sequence of tetrahedral Si, Al, O layers and

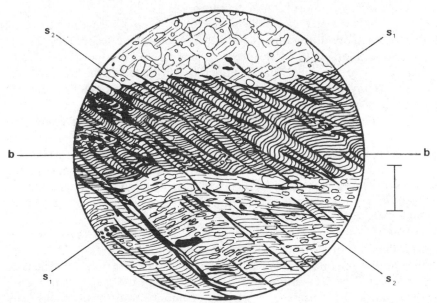

Figure 5.16 This section of graphitic phyllite interbedded with sandy slate, Lotschental,
Canton Valais, Switzerland. Scale bar 1 mm

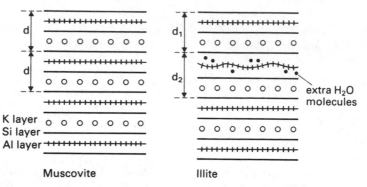

Figure 5.17   Schematic sketch of the structure of muscovite (left) and illite (right), looking along the layers of silicon and aluminium atoms. The d-spacings of the layers are shown. Illite sometimes contains additional H_2O atoms between the Al and Si layers, making some d-spacings (d_2) larger than usual.

octahedral Al, O layers as muscovite (Deer *et al.* 1966), while others have additional layers of H_2O, as in the clay mineral illite. As the metamorphic grade increases, the number of H_2O layers decreases, and this change can be expressed as a decrease in the relative proportion of illite compared with muscovite in the white mica (Fig. 5.17). This is a type of continuous metamorphic reaction involving loss of H_2O with increasing temperature, and it has already been shown that such dehydration reactions are characteristic of the progressive metamorphism of pelitic rocks.

The progressive change of the white micas from illite to muscovite is followed by X-ray diffraction and electron microscopic techniques. At the highest resolution of the transmission electron microscope (**TEM**), individual layers of the sheet silicate minerals can be distinguished, and the thicker spacing between illite layers with additional H_2O can be directly recognized.

X-ray diffraction is easier to carry out, and here the progressive disappearance of the wider-spaced illite layers, resulting in a higher proportion of regularly spaced muscovite layers, leads to a sharpening of the peaks due to diffraction from the muscovite layers (Fig. 5.18). The sharpness of the peaks is measured as the ratio of their width to height, and a number known as the **illite crystallinity** is derived from these ratios (Dunoyer de Segonzac 1970).

In the white micas of unmetamorphosed shales the illite crystallinity is greater than 7.5, in white micas from the stilpnomelane zone it is 4.0–7.5 and in the chlorite zone of higher metamorphic grade it is less than 4.0. The illite crystallinity of the rock in Figure 5.16 is 3.2. That of the specimen in Figure 5.14 has not been determined directly, but comparison with specimens from the nearby Linth Valley suggests that the value should lie in the range 5.5–6.0 (Frey & Hunziker 1973).

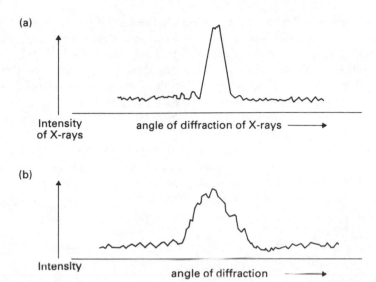

Figure 5.18 The peaks of intense diffraction of X-rays from powders of muscovite (a) and illite (b). Because there are some layers with a different spacing in illite, the peaks are broader. Illite crystallinity numbers represent this broadening effect.

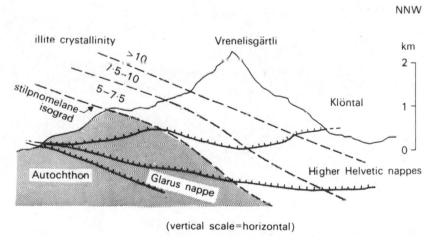

(vertical scale=horizontal)

Figure 5.19 Cross section along the western side of the Linth Valley, Canton Glarus, Switzerland, showing the variation of illite crystallinity with height, and the incoming of stilpnomelane. The surface marking the first appearance of stilpnomelane is called an isograd.

In this area there is a vertical relief of more than 3000 m, so it is possible to determine the shape of the **isograd** surfaces in three dimensions. The isograds dip northwards, and are approximately parallel to the upper surface of an uplifted body of pre-Alpine crystalline rocks, the Aar Massif. Figure 5.19 shows the form of the isograd surfaces in cross section along the western side of the Linth Valley. The mineral assemblages and the illite crystallinity values correlate well with one another as indicators of metamorphic grade. Several rocks from the area have also been dated radiometrically by the potassium–argon method, applied to glauconite, stilpnomelane and biotite. The resulting dates are similar at all metamorphic grades, implying that the dates concerned are those of the heating episode which imprinted the metamorphic pattern. In many regional metamorphic terrains, it is thought that isograds have the arched form shown in Figure 5.19, but it can seldom be so convincingly demonstrated.

The isograds cut across the thrust planes between nappes, demonstrating that heating took place after the thickening of the sedimentary sequence by thrusting, and that heat was supplied from below. This pattern accords well with thermotectonic models for metamorphism in areas of plate convergence (Chapter 6).

6 Metamorphism in subduction zones

Subduction zones are associated with the greatest disturbances to the flow of heat from the interior of the Earth. They are also the sites of large departures from isostatic equilibrium in the lithosphere, particularly in the regions where oceanic plates bend to begin their descent into the Earth's mantle.

As discussed in Chapter 5, oceanic crust cools progressively as it spreads away from oceanic ridges, and therefore in most subduction zones, which are distant from oceanic ridges, the oceanic lithosphere is cool. This cooling results in unusually low geothermal gradients on the oceanward side of subduction zones (except in the unusual sites where an oceanic ridge is undergoing subduction). By contrast, the lithospheric wedge above the subduction zone includes a region of high heat flow, associated with the volcanic arc of the subduction zone. Figure 6.1 shows the pattern of isotherms in a subduction zone, calculated by thermal modelling, assuming that heat is lost by conduction from the descending lithospheric slab. This diagram reveals a number of significant features of the thermal structure of subduction zones.

The descending slab heats up slowly compared with its rate of descent into the mantle, and therefore the rocks within it remain relatively cool to very

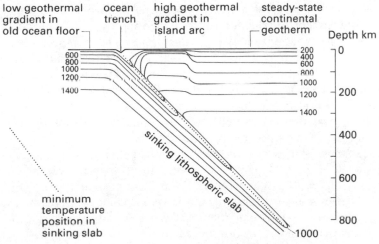

Figure 6.1   Thermal structure of a subduction zone, from Bott (1982).

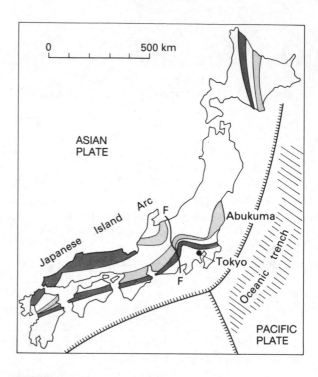

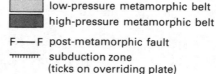

Figure 6.2 Map of the metamorphic belts of the Japanese Islands, and the three subduction zones which meet nearby.

low-pressure metamorphic belt
high-pressure metamorphic belt
F——F	post-metamorphic fault
ᴛᴛᴛᴛᴛᴛ	subduction zone (ticks on overriding plate)

considerable depths, greater than the depth to the Moho in ordinary continental lithosphere. Subduction zones therefore provide suitable conditions for metamorphism at unusually high pressures and low temperatures. There is a peculiar type of metamorphic rocks called **blueschists**, which were metamorphosed under just such conditions in the **blueschist facies**, and they occur in regions in which there is independent evidence for the existence of subduction zones at the time of metamorphism. The place where this relationship was first discovered was Japan, and it has subsequently been found in many other ancient subduction zones.

The Japanese islands lie above two presently active subduction zones, and many of the rocks of Japan were formed in geologically recent times by volcanism and uplift associated with the present day subduction (Fig. 6.2). Beneath these younger rocks, there are rocks which were formed during earlier episodes of subduction, including well marked periods in Triassic to Jurassic times, and in Cretaceous times. During these episodes, basic igneous rocks derived from the ancient floor of the Pacific Ocean, and sedimentary rocks scraped from the surface to form accretionary complexes, underwent metamorphism, giving rise to pairs of belts of metamorphic rocks. The two belts of each pair consist of a belt of rocks which underwent regional metamorphism at relatively high temperatures and low pressures (high T–low P) on the side nearer the Asian continent, and a belt of rocks where the metamorphism occurred at relatively low temperatures and high pressures (low T–high P) on the Pacific Ocean side. It is the ocean-side low T–high P belt which includes the blueschists.

The belts can be identified with some confidence with parts of the subduction zone shown in Figure 6.1. The high T–low P belt comes from the region in the over-riding wedge where there is active volcanism, and the geothermal gradient is relatively high. The low T–high P belt represents material derived from the descending lithospheric plate, probably scraped from its upper surface to constitute the deeper parts of the accretionary complex. The occurrence of blueschists fits in with this idea of their origin well, because they are frequently found as blocks in **tectonic melange**.

Figure 6.3 shows depth–temperature–time paths for rocks in the descending

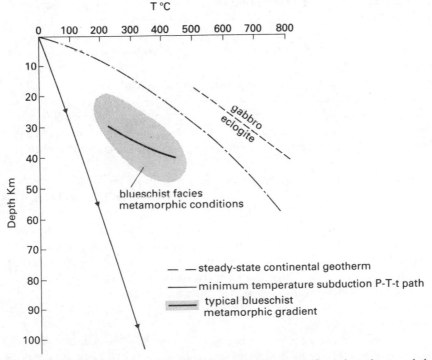

Figure 6.3 Depth–temperature plot showing the steady-state continental geotherm and the curve of *minimum* temperature at each depth within the subducting lithospheric slab. P–T conditions of the blueschist facies are shown, with a typical blueschist metamorphic field gradient.

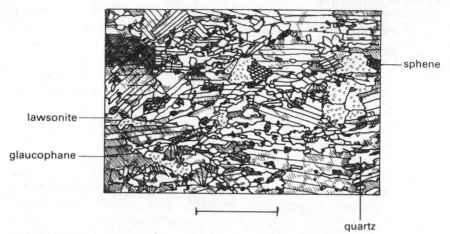

Figure 6.4 Glaucophane–lawsonite schist from Berkeley Hills, California, USA. Scale bar 1 mm.

oceanic lithospheric plate, once more derived from simple calculations based on thermal conduction. Because most of the plate descends to depths as great as 700 km in the mantle, unlike the P–T–t paths we have seen before, these paths do not return to low temperatures and surface pressures. Also shown on the diagram is the field of P–T conditions of blueschist metamorphism, estimated by studying the mineral assemblages of blueschists. The diagram makes it very clear that conditions in the descending lithospheric plate would certainly be suitable for blueschist metamorphism, and the tectonic problem is one of proposing a mechanism which can return the blueschists to the surface, rather than subducting them down deep into the mantle.

Figure 6.5 shows a thin section of a schist of basic igneous composition, not from Japan, but from the Franciscan Complex of California, USA. This is of comparable age to the Sanbagawa low T–high P metamorphic belt of Japan, but in California subduction of the Pacific Plate has been succeeded by transcurrent displacement on **transform faults** such as the San Andreas Fault. This has jumbled the metamorphic belts, so the regular arrangement as seen in Japan is not apparent. The Franciscan Complex includes basic rocks of ocean-floor character, and sedimentary rocks which are probably derived from an accretionary prism. The basic igneous rocks occasionally show pillow structures.

The rock is called a blueschist because in hand specimen, and when seen under low magnification down the microscope, it is coloured blue by the presence of a sodium-rich amphibole, glaucophane. The basic igneous rock contains this amphibole, rather than the actinolite or hornblende which would form during lower pressure regional metamorphism, because high pressure favours the coupled replacement of Ca and Mg in amphibole by Na and Al:

$$Ca^{+2}Mg^{+2} = Na^{+1}Al^{+3}.$$

In amphiboles, over the temperature range of blueschist metamorphism, this results in a change from actinolite to glaucophane:

$$Ca_2Mg_5[Si_8O_{22}] (OH)_2 = Na_2Mg_3Al_2[Si_8O_{22}] (OH)_2$$

$$\text{actinolite} = \text{glaucophane}$$

The glaucophane has a prismatic crystal habit, and the long axes of the crystals lie in one plane giving the rock a penetrative foliation. Basal sections through the prisms show the characteristic {110} cleavages making an angle of 120° to one another. The blue colour of some sections in plane polarized light is not in itself diagnostic, but the total pleochroic scheme of the amphibole does identify glaucophane. It is: α – colourless, β – lavender-blue, γ – blue. The extinction angle γ: z is up to 15° (lower than hornblende), the optic sign is negative, with a moderate to low 2V.

The other essential mineral in the sample is lawsonite, $CaAl_2(OH)_2[Si_2O_7]H_2O$. This is a hydrated silicate of calcium and aluminium, whose chemical formula is equivalent to that of anorthite, $CaAl_2Si_2O_8$ plus $2H_2O$, and thus lawsonite has taken the place of the anorthite component in the plagioclase of the parent basaltic rock. It is orthorhombic and optically positive, which distinguishes it from epidote (which is monoclinic and negative). The two minerals are frequently confused, and may occur together in the same rock, as they do in Figure 6.4. Unlike glaucophane, lawsonite is not a mineral whose temperature range of stability is increased by increasing pressure, but it is only stable at comparatively low temperatures (up to about 350°C). The co-existence of glaucophane and lawsonite in rocks of basic igneous composition therefore of itself suggests low T–high P metamorphic conditions.

Like most blueschists, this rock shows a variety of textures in the metamorphic minerals. The blue amphiboles have a strong preferred orientation, and are comparatively uniform in size, but study under the high power of the microscope shows that the colour varies from the core to the rim of the amphiboles. The change in colour reflects a change in chemical composition, i.e. the glaucophanitic amphiboles are zoned. The lawsonite crystals also display a preferred orientation. The plagioclase and quartz crystals show fragmentation and smaller crystals developing along grain boundaries.

This kind of textural evidence for disequilibrium is a frequent feature of blueschists, and their field relationships are often also confused, with patches of medium- to coarse-grained blueschists occurring among metamorphosed

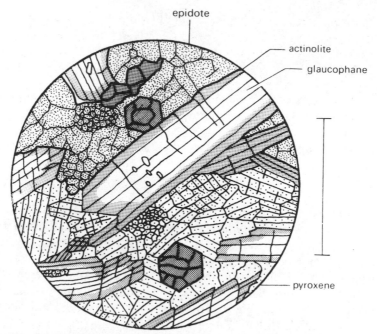

Figure 6.5 Glaucophane–epidote schist, Riffelhaus, near Zermatt, Switzerland. Harker thin section collection, Cambridge University, no. 19435. Scale bar 1 mm.

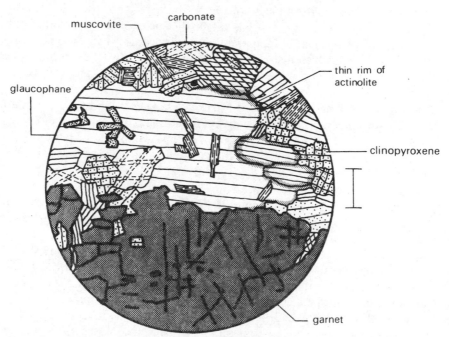

Figure 6.6 Glaucophanitic eclogite, Pfulwe, Zermatt, Switzerland. Harker thin section collection, Cambridge University, no. 92882. Scale bar 1 mm.

greywackes apparently showing rather uniform low-grade metamorphism. Progressive metamorphic sequences are seldom found in which isograds can be mapped in the field, although in some parts of the Franciscan Complex isograds have been mapped by comparing large numbers of thin sections.

Figure 6.5 shows a coarser-grained blueschist, which comes from the Alps in Switzerland. In the Alps, blueschists have survived from episodes of subduction which occurred along the flanks of an ocean called Tethys, which was destroyed later by a collision of the continental plates on either side. Some of the rocks of the Alps have mineral assemblages which record the highest pressures of metamorphism which have so far been discovered in crustal rocks (Chopin 1984). The rocks of Figures 6.5 and 6.6 come from the same tectonic zone as the very high-pressure rocks, but from a part where pressures were lower. The zone lies south and southeast of the sedimentary basins described in Chapter 5, and major deformation and metamorphism took place at an earlier date, at 70–80 Ma. The metamorphic rocks lie in a pile of thrust sheets or nappes, and are called the Pennide Nappes.

The metamorphic sequences of the Pennide Nappes are varied in their original composition. Pelitic rocks are quite common, but the blueschist mineral assemblages are best developed in rocks of basic igneous composition. Features such as pillow lava structure suggest that some of these basic rocks are fragments of ophiolite. The rock in Figure 6.5 differs from that of Figure 6.4 because it contains metamorphic pyroxene. It is a sodium-rich variety of augite known as **omphacite**, and its optical properties are like those of augite except that the omphacite has a pale-green colour in plane-polarized light. The rock also contains subhedral to euhedral garnets. They have a relatively high proportion of the magnesium garnet component, pyrope $(Mg_3Al_2Si_3O_{12})$ compared with the iron garnet, almandine $(Fe_3Al_2Si_3O_{12})$, which is higher in the garnets described in Chapter 3 and Chapter 7. The planar crystal boundaries, and euhedral outlines of the crystals might suggest that the rock has an equilibrium mineral assemblage, but the amphiboles in the rock show that this is not the case. Under the petrological microscope in plane polarized light the prismatic amphiboles can be seen to be zoned, their cores showing the blue and lavender colours of glaucophane, their rims the dark green of actinolite. Between crossed polars it can also be seen that the birefringence of the rims is higher than the core.

It is not easy to work out the equilibrium mineral assemblage (or assemblages) for this rock. One possible interpretation is that there was an early assemblage garnet + omphacite + glaucophane + epidote, which later broke down locally to albite + actinolite (there is a small rim of albite in the upper left of Figure 6.5). If this is correct, the early assemblage would be a transitional one between blueschist and **eclogite**, which is a pyrope-rich garnet + omphacite rock (Chapter 9). The later assemblage crystallized at a stage on the P–T–t path when the pressure had fallen from the high values in the subduction zone.

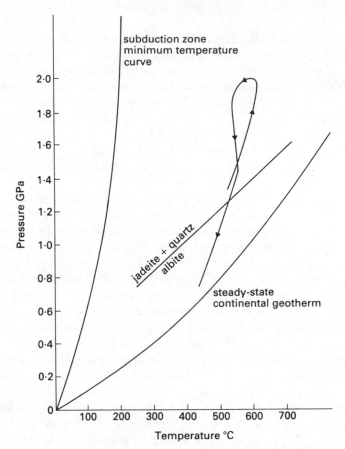

Figure 6.7 P–T–t path for blueschists and eclogites from the Zermatt region of Switzerland, with steady-state continental geotherm and subduction zone minimum temperature curve, as in Figure 6.3.

Figure 6.6 shows a rock which displays three episodes of metamorphism on its P–T–t path. A large garnet crystal is in contact with omphacite, which appears to be undergoing replacement by a large glaucophane crystal associated with a carbonite mineral, probably calcite or dolomite. Small prisms of epidote and flakes of muscovite are growing inside the glaucophane crystal, which has a narrow rim of actinolite, similar to the crystals in Figure 6.5.

In this rock, there appear to have been three successive mineral assemblages:

high P   garnet + omphacite

↓

glaucophane + calcite

↓

low P   actinolite + muscovite + epidote

All the blueschists we have seen do not show a single mineral assemblage, but there are features such as mineral zoning, which were described in Chapter 2 as indicators of disequilibrium. It has been possible to speculate on their P–T–t history by identifying successive mineral assemblages, regarded as having evolved in small domains of equilibrium which existed at different times in the rocks' history. The field relationships of blueschists are similarly complicated. They often occur in **melanges**, which consist of blocks of different rocks of different sizes, shapes and rock types. The blocks are often cut by fractures or extended into **boudins**, and are separated by a matrix either of turbidite sediments, or of serpentinite. Some blocks are exotic, shallow-water sediments among deep-water sediments, or sediments among serpentine. Blueschists are often found as blocks in melange, and in some cases large blocks show zoning on a scale of metres, with blueschist metamorphic assemblages in the cores of the blocks, and assemblages such as albite + epidote + actinolite (**greenschists**) round their margins.

The mineral assemblages define the P–T–t loop shown in Figure 6.7. It shows early stages when pressure rose while temperatures stayed low, as the rock was carried down the subduction zone, followed by an increase in temperature with falling pressures. The later stages of the P–T–t loop could be explained if part of the subducting plate became uncoupled from the rest of the plate, and stuck to the overriding plate. If this process happened for a succession of slices from the oceanic sediments and basic volcanics from the top of the descending plate, the overriding plate would grow by accretion of the material scraped from the lower plate. Such accretionary complexes can be identified in present-day subduction zones, and the deeper levels within them would consist of a melange of sediments with basic igneous rocks, including blueschists.

The rocks of the Sanbagawa metamorphic belt of Japan, and the Franciscan Complex of California frequently show melange structure, supporting the theory that they formed in ancient subduction zones. The melange described in Chapter 4 from Turkey contains glaucophane (Hall 1980), and was probably also associated with a subduction zone.

7 Metamorphism in collision zones

Chapter 6 described the characteristic types of metamorphism which occur around subduction zones, where oceanic crust is thrust downwards into the mantle. This process of destruction of oceanic crust will eventually change or end, by changes in the distribution of plates (e.g. when an oceanic ridge is subducted) or by the arrival of continental crust at the subduction zone in the overridden oceanic plate. When such continental fragments are small, they may be incorporated into the evolving orogenic belt above the subduction zone as allochthonous terranes. When the continental fragments are large, there will be a continental collision. Continental collisions of this type are occurring at the present day in the Himalayan mountain chain, where the Indian Plate is colliding with the Asian Plate, and in the Middle East, where the Arabian Plate is colliding with the Asian Plate. The existence of mountain chains formed by continental collision within what are now stable continental plates (e.g. the Ural Mountains of the USSR) shows that eventually such collisions 'lock' subduction zones.

The sequence of events from the rifting apart of continents to form oceans, to the stabilization of continental crust after a continental collision may be summarized in the **Wilson Cycle** (Fig. 7.1), named for the geophysicist J. Tuzo Wilson, who played a large part in establishing this theory for the evolution of mountain belts. The cycle summarizes the events in the evolution of collision orogenic belts such as the Himalayas, and the stages in it do not necessarily follow inevitably one after another. In some cases the stages were separated by many millions of years when the part of the lithosphere involved was relatively stable, and recent tectonic research has demonstrated the complexity of the later, collisional stages 3 and 4, because of the involvement of allochthonous terranes. Bearing these reservations in mind, however, the Wilson Cycle is a convenient summary of the sequence of processes in plate collision zones, and will be used as a framework for the discussion in this chapter.

The effect of lithospheric collision is to thicken the lithosphere, especially the upper layer of continental crust. This is because continental crust is less dense than oceanic crust, and resists subduction into the mantle because of its buoyancy, although occasionally it is subducted (Chopin 1984). The presence of thickened continental crust in areas of collision such as the Himalayas, and former collision zones, such as the Alps, has been established by geophysical methods. Two models for the process of thickening

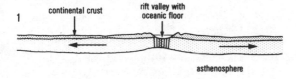

1 continental crust   rift valley with
                       oceanic floor

asthenosphere

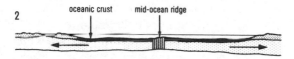

2   oceanic crust   mid-ocean ridge

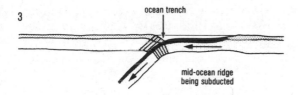

3   ocean trench

mid-ocean ridge
being subducted

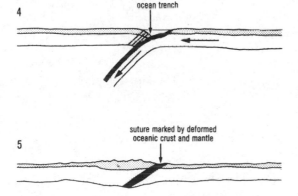

4   ocean trench

Figure 7.1 Wilson Cycle for the formation of orogenic belts by extension leading to ocean formation, followed by continental collision. 1. Onset of extension (Red Sea stage). 2. Maximum extension (Atlantic Ocean stage). 3. Subduction of wide ocean floor (Pacific Ocean stage). 4. Subduction of narrow ocean with continental fragments (Mediterranean stage). 5. Continental collision (Himalayan stage).

5   suture marked by deformed
     oceanic crust and mantle

☐ continental crust   ☐ lithospheric mantle

will be considered here – thickening by homogeneous strain of the lithosphere (Model I), and thickening by overthrusting (Model II). Model I is more likely to be followed when rates of collision are slow and if the lithosphere is plastic, Model II when rates of collision are high and if the lithosphere is brittle. The two models are not mutually exclusive, but rather two extremes in a spectrum of possibilities.

Model I. Homogeneous strain of the lithosphere

If a prism of lithosphere is caught between two converging plates, it may deform plastically, shortening in a horizontal direction, at right angles to the direction of convergence of the plates, and extending in a vertical direction, so that it becomes thicker. The deformation is assumed to take place by pure shear, with the maximum compression direction horizontal, and the maximum extension vertical (Fig. 7.2). It is assumed that there is no extension or compression in the horizontal direction normal to the direction of convergence (i.e. in and out of the plane of section of Fig. 7.2). In both Model I and Model II it is assumed that rates of deformation are much quicker than rates of heating or cooling of thick slabs of lithosphere, and also, to simplify discussion, that the tectonic processes of thickening and later isostatic uplift take place in separate stages. Most published models (e.g. Thompson 1983) which are more realistic in assuming overlaps in time between collision, uplift, and thermal relaxation, yield similar results to the present discussion. While the present simplifications are not serious for purposes of general discussion, more accurate models should be used for quantitative analysis of metamorphic histories.

Because deformation is more rapid than thermal relaxation, the isotherms in the lithosphere are shown deforming as passive strain markers in the prism of plastic lithosphere. Thus the initial collision produces a perturbed geothermal gradient which is lower than the initial gradient (Fig. 7.2 a & d). This relaxes back to a steady-state geothermal gradient (Fig. 7.2 d), with rapid uplift due to isostatic adjustment after collision tending to produce high geothermal gradients near the surface for a time (Fig. 7.2 c).

Most of the rocks within the prism will follow clockwise P–T–t paths on a P–T plot which will be anti-clockwise on a depth–temperature plot (Fig. 7.2 e, f).

Model II. Lithospheric thickening by overthrusting

Like Model I, this model assumes that thickening and uplift are much quicker than heating and cooling. The rigid lithosphere is taken to have shortened by overthrusting involving **ramps**, cutting obliquely upwards through the major layers of the lithosphere, and **flats**, running along the horizontal discontinuities in the lithosphere. The actual model presented here is a very simple one, assuming that one ramp cuts upwards through the lithospheric mantle, there is a flat at the base of the crust at the **Moho discontinuity**, and another ramp cutting up to the surface through the entire thickness of the continental crust (Fig. 7.3). While actual collision belts such as the Himalayas have a more complicated thrust pattern than the model, they do have important flats at the Moho, and the model is a reasonable one for general discussion purposes.

A feature of all thrusting models, including this one, is that immediately

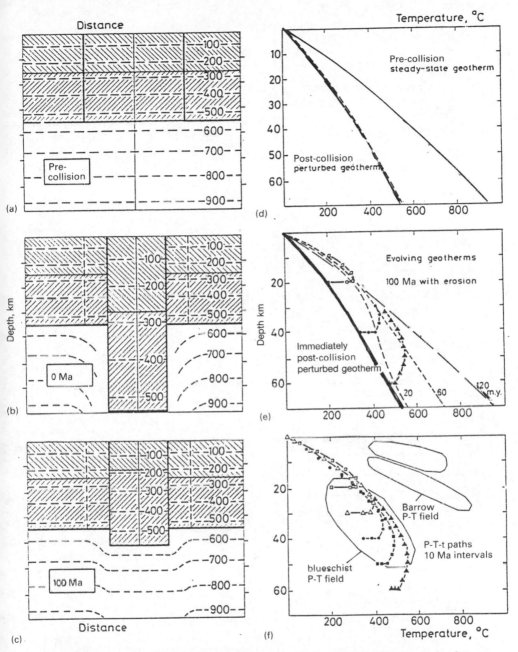

Figure 7.2 Thermotectonic model for continental collision by homogeneous strain, from Thompson (1981). (a) Steady-state conditions at the onset of collision. (b) Deformation of prism of continental lithosphere by homogeneous strain after collision. The isotherms have moved further apart, producing a perturbed geotherm with a low geothermal gradient. (c) 100 Ma after collision, with uplift and erosion of deformed orogenic prism, and partial relaxation of isotherms. (d) Post collision perturbed geotherm compared with steady-state geotherm. (e) Evolving geotherms, at 20, 60 and 120 Ma after collision. (f) Depth–T–t curves for rocks at different depths within the deformed prism.

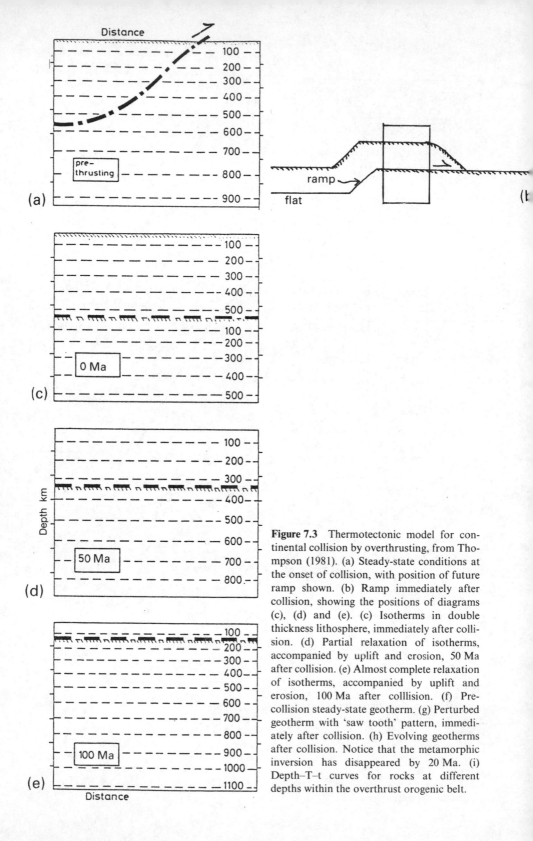

Figure 7.3 Thermotectonic model for continental collision by overthrusting, from Thompson (1981). (a) Steady-state conditions at the onset of collision, with position of future ramp shown. (b) Ramp immediately after collision, showing the positions of diagrams (c), (d) and (e). (c) Isotherms in double thickness lithosphere, immediately after collision. (d) Partial relaxation of isotherms, accompanied by uplift and erosion, 50 Ma after collision. (e) Almost complete relaxation of isotherms, accompanied by uplift and erosion, 100 Ma after colllision. (f) Precollision steady-state geotherm. (g) Perturbed geotherm with 'saw tooth' pattern, immediately after collision. (h) Evolving geotherms after collision. Notice that the metamorphic inversion has disappeared by 20 Ma. (i) Depth–T–t curves for rocks at different depths within the overthrust orogenic belt.

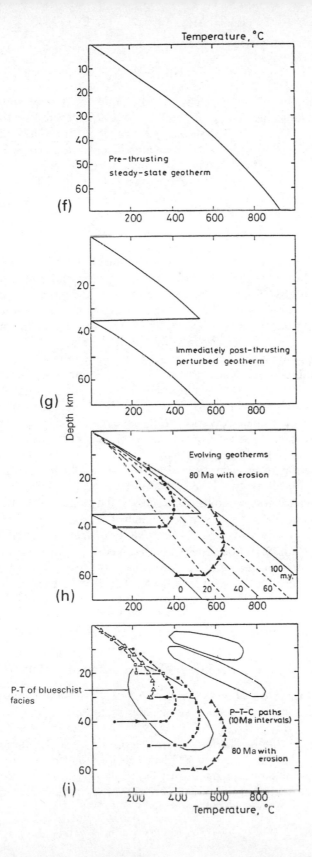

Temperature, °C

(f) Pre-thrusting steady-state geotherm

(g) Immediately post-thrusting perturbed geotherm

Depth km

(h) Evolving geotherms
80 Ma with erosion
100 m.y.
0   20   40   60

(i) P-T of blueschist facies
P-T-C paths (10 Ma intervals)
80 Ma with erosion

Temperature, °C

A feature of all thrusting models, including this one, is that immediately after overthrusting, high *inverted* geothermal gradients are present at the thrust surfaces (Oxburgh & Turcotte 1974) (Fig. 7.3 g). They disappear relatively quickly as the geothermal gradient relaxes towards the steady-state gradient (Fig. 7.3 h). The gradient takes much longer to relax from its low average value after collision to a more normal value (Fig. 7.3 h & i).

Most P–T–t paths, as in Model I, have a clockwise form on a P–T graph, with exceptions being found near the overthrust surfaces, and near the ground surface, where as in Model I, rapid uplift perturbs geothermal gradients (Fig. 7.3 i).

Both models therefore predict that regional metamorphism in collisional zones should show an initial episode of metamorphism under high pressures at low temperatures, followed by lower pressures at higher temperatures. The overthrust model also predicts that inverted metamorphic gradients may be preserved in some cases. Two examples of areas of regional metamorphic rocks will now be discussed, the first illustrating the clockwise P–T–t paths, the second showing an inverted metamorphic field gradient. The classic Barrow's metamorphic zones of the Grampian Highlands of Scotland, mentioned in Chapter 1, will not be used here because although this area does show clockwise P–T–t paths (Dempster 1986) and probably inverted metamorphic field gradients (Watkins 1983), the relationship between tectonic and metamorphic processes in the area remains controversial and the extent of the inverted field gradient is not known.

But the Caledonian orogenic belt of northwestern Europe is a good area for illustrating the Barrow type of metamorphism, and showing that it is related in a general way to collision processes. The chain is the northeastward extension of the Appalachian orogenic belt of eastern North America, and was continuous with it before the Atlantic opened about 80 Ma ago. Much earlier than that, the Appalachian–Caledonian orogenic belt was formed by a complex sequence of collisional processes between 500 and 420 Ma ago (Wilson 1966, Windley 1984). The British part of the chain shows the results of collisions of allochthonous terranes with the Canadian Shield, mainly by strike–slip motion (Mason 1988), the Scandinavian part shows terranes overthrust above the Baltic Shield (Stephens 1988).

Metamorphic evolution of pelitic rocks of Connemara, Ireland

The rocks of the Connemara district of western Ireland belong to an allochthonous terrane with many similarities to the Grampian Terrane of Scotland but in a distinctly different position in the Caledonian orogenic belt (Hutton & Dewey 1986). The terrane was emplaced by strike motion along bounding transform faults about 430 Ma ago, and had earlier undergone deformation and metamorphism during collision, probably with the Canadian Shield. The metamorphic rocks were deposited in later Proterozoic to

Cambrian times, and underwent collisional deformation (as shown by the major and minor tectonic structures) and metamorphism about 490 Ma ago. Like the metamorphic rocks of the Grampian terrane, they are assigned to the Dalradian Supergroup, and include quartzites, schists, marbles and amphibolites. The Dalradian rocks are buried beneath unmetamorphosed Ordovician, Silurian and younger rocks (Fig. 7.4). There is an increase in metamorphic grade from north to south across the terrane.

Figure 7.5 shows a schist from the Dalradian of Connemara. It has the mineral assemblage: cordierite + quartz + sillimanite + biotite + plagioclase + opaques + apatite. Sillimanite is a prismatic mineral with the formula Al_2SiO_5, and is one of the index minerals of the Grampian metamorphic zones. It is one of three minerals (the other two are andalusite and kyanite) with this formula. This mineral assemblage resembles those of the high temperature hornfelses in the Skiddaw aureole, except that sillimanite replaces andalusite. Study of the Al_2SiO_5 phase diagram suggests that this is because the Connemara schist crystallized under higher pressures than the Skiddaw hornfelses. The sillimanite in Fig. 7.5 consists of small needles, often running sub-parallel to the cleavage planes of adjacent flakes of biotite. This textural variety of sillimanite is called **fibrolite**. A drawing of the same rock under a higher power of magnification (Fig. 7.6) shows that the sillimanite needles occur near the grain boundaries between biotite and quartz

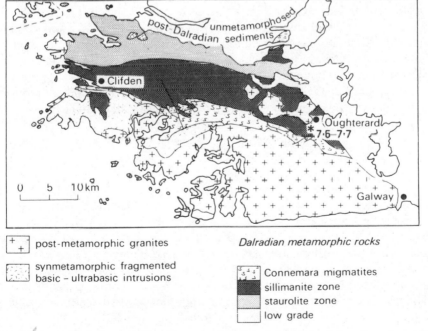

+ +	post-metamorphic granites		*Dalradian metamorphic rocks*

synmetamorphic fragmented
basic – ultrabasic intrusions

Connemara migmatites
sillimanite zone
staurolite zone
low grade

Figure 7.4 Geological map of the Connemara District, western Ireland. Location of samples shown in Figures 7.5 to 7.7 shown.

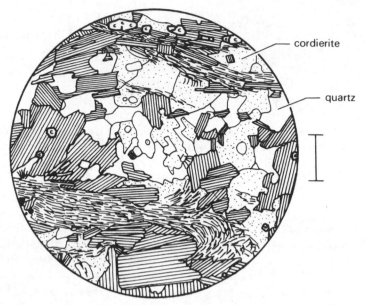

Figure 7.5 Cordierite–sillimanite schist, Oughterard, County Galway, Ireland. Scale bar 1 mm.

or cordierite, and penetrate into both those minerals. The interpretation of this texture is that sillimanite crystals **nucleated** (i.e. began to grow) at the grain boundary, and grew into the other minerals. The implication is clearly that the sillimanite grew later than the biotite, quartz and cordierite. A close study of the sillimanite prisms shows that they become thicker away from the pre-existing grain boundaries, suggesting that they grew over an appreciable period of time, as the temperature was rising, making them coarser-grained. Since sillimanite is the high temperature polymorph of Al_2SiO_5, this implies that the temperature rose during this period.

Figure 7.7 shows another pelitic rock from Connemara containing garnet, quartz, cordierite, plagioclase, opaques and apatite. It is not clear whether this list of minerals is an equilibrium mineral assemblage, because the garnet is present in small isolated volumes, separated by cordierite from biotite (so all the minerals do not show mutual grain boundaries). It appears that the garnet is a **relict mineral**, which was outside the P–T conditions of its stability when the other minerals in the rock crystallized. The lower half of the field shown in the drawing represents the remains of a porphyroblast of garnet, which has broken down to cordierite, biotite and sillimanite. This cannot be a simple reaction, because biotite contains potassium, while garnet does not. The reaction may be balanced by assuming that potassium was introduced into the garnet as ions (K^+) along with metamorphic fluid, mostly H_2O:

$$garnet + H_2O + K^+ = cordierite + biotite + sillimanite$$

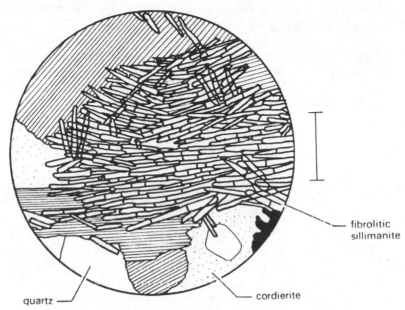

Figure 7.6   Enlarged drawing of part of the rock shown in Figure 7.5. Scale bar 0.1 mm.

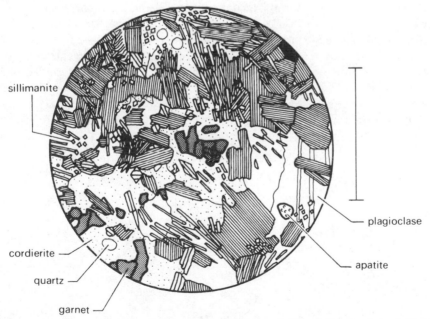

Figure 7.7   Garnet–cordierite–sillimanite schist, Oughterard, County Galway, Ireland. Scale bar 1 mm.

The source of the K^+ ions may well have been in the groundmass surrounding the garnet porphyroblasts, because this contains a relatively high proportion of biotite. The potassium ions were probably set free by reactions between phyllosilicate minerals. The groundmass now contains no chlorite and muscovite, and if these were present in the early high pressure–low temperature phase of metamorphism, K^+ would be released by the biotite-forming reaction:

$$\text{chlorite} + \text{muscovite} = \text{biotite} + H_2O + K^+$$

The breakdown reactions of garnet have been investigated experimentally, and the reaction given would require an increase in temperature and a decrease in pressure to go from left to right, supporting the clockwise shape of the P–T–t paths for Connemara. A representative P–T–t path for Connemara, with some mineral stability fields indicated, is shown in Figure 7.8.

The pattern of metamorphic zones in this area shows a variation from metamorphism under low temperatures in the north, to high temperatures in the south (Fig. 7.4). From the discussion in the earlier parts of this book, it is likely that most of the metamorphic mineral assemblages in the rocks crystallized under the highest temperatures reached during the P–T–t cycles, and therefore that the pattern of metamorphic zones reflects the conditions in the Earth's crust during the later stage in the metamorphism. In the Connemara area, the source of the heat for this metamorphism has been discovered. There are intrusions of basic and ultrabasic igneous rocks in the south of the area, always found near the highest zones of metamorphism. Between the intrusions and the Dalradian metamorphic rocks there is often a zone of granitic gneisses, with layers of granite on a mm to cm scale, separated by layers of high-grade schists. These are known as **migmatites**, and the origin of such rocks will be discussed in more detail in Chapter 8. Looking ahead, we can say that the migmatites represent a level in the crust

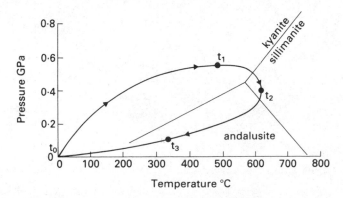

Figure 7.8 P–T–t path for the rocks from Oughterard, Connemara.

where the large quantities of heat released from the cooling basic and ultrabasic igneous rocks melted the Dalradian rocks. Thus the later metamorphic episode at Connemara shows that geothermal gradients were rather high at this time, during the later stages of the Caledonian collision. The reason for the high temperature gradients was the intrusion of large volumes of magma into the crust. This is not a mechanism which is included in either Model I or Model II, which would thus have to be somewhat modified for the Connemara case.

The Sulitjelma copper-mining district, Norway and Sweden

The Caledonian orogenic belt in Scandinavia is different from that in the British Isles because the allochthonous terranes were thrust over the Baltic Shield as relatively thin sheets, or nappes, and have not subsequently been dismembered by later movements on near-vertical faults (Stephens 1988, Mason 1988). The thrusting took place between 440 and 420 Ma ago, and much of the regional metamorphism occurred just before thrusting, or at the same time (Burton et al. 1989, Barker & Anderson 1989). In the Sulitjelma District, a pattern of metamorphic zones is preserved, which apparently formed during and after a major episode of thrusting.

The Sulitjelma copper-mining district lies just north of the Arctic Circle, and straddles the international boundary between Norway and Sweden. The allochthonous metamorphic rocks were thrust southeastwards over the Baltic Shield during a continental collision about 430 Ma ago (Stephens 1988). The rocks of Sulitjelma lie in a relatively high position in the pile of thrust-sheets or nappes, and the copper ores, like those of Cyprus (Chapter 5) lie at or near the upper surface of an ophiolite complex (Boyle 1985, Mason 1985). The ophiolite complex represents crust with the characteristic ocean-floor structure, formed by extension before the overthrusting. But petrological and geochemical study of the ophiolite shows that it is probably crust from an oceanic basin above a subduction zone, rather than from an oceanic ridge. Unlike the Troodos Ophiolite of Cyprus, the Sulitjelma Ophiolite was buried beneath a thick sequence of clastic sedimentary rocks. Scarce fossils in the eastern part of the area suggest that they are of Ordovician to Silurian age. The sedimentary sequence consists mainly of rocks deposited by turbidity currents in an unstable sedimentary basin (Kirk & Mason 1984), known as **flysch**. The clastic sediments contain an appreciable proportion of volcanic material derived from the island arc.

The rocks of the Sulitjelma Ophiolite have undergone regional metamorphism, which occurred during the continental collision, and has almost completely destroyed the metamorphic assemblages which probably formed during oceanic extension. They are called the Sulitjelma amphibolites, and the metamorphosed turbidites which originally buried them are known as

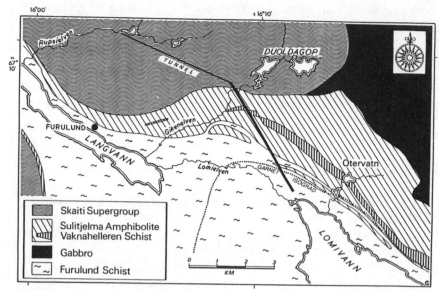

Figure 7.9 Geological map of part of the Sulitjelma copper-mining district, Norway, showing the location of the hydroelectric tunnel of Figure 7.10. The Gabbro and Sulitjelma Amphibolites are parts of the Sulitjelma Ophiolite.

the Furulund Schist Group, or Furulund schist for short. The overthrusting caused overturning of the ophiolite and its sedimentary cover, so that the Sulitjelma amphibolites occur *above* the Furulund schist. Both rock sequences show a variable grade of regional metamorphism, high in the west, and lower in the east. The Sulitjelma amphibolites and Furulund schist can be mapped continuously from low-grade to high-grade areas, and individual stratigraphical sub-divisions within them can be identified, crossing metamorphic boundaries (Fig. 7.9). This feature makes the Sulitjelma district a good one for studying the pattern of metamorphic zones due to regional metamorphism, compared with other areas (such as Barrow's Zones in the Grampian Highlands of Scotland) where stratigraphic boundaries and metamorphic zone boundaries are parallel to one another, so there is uncertainty whether changes in the metamorphic mineral assemblages are due to changes in metamorphic grade or to changes in rock composition.

Progressive regional metamorphism of the Furulund Schist Group

The Furulund schist consists predominantly of phyllosilicate minerals such as muscovite, biotite and chlorite. They show a planar foliation fabric, which is usually parallel to the sedimentary bedding, although local discordances can be found (Kirk & Mason 1984). They also contain quartz and plagioclase feldspar, and usually an appreciable proportion of a carbonate mineral, calcite, dolomite or ankerite. There are accessory amounts of the opaque phases pyrite, chalcopyrite and ilmenite, and the transparent minerals sphene, apatite and minerals of the epidote group. In addition some samples contain garnet and prismatic or needle-shaped amphibole minerals.

The original bedding of the Furulund schist is seen in outcrop as a darker or lighter green or brown coloration. The quartz and plagioclase gains often retain their primary sedimentary size, and their sedimentary shape, a little modified by metamorphic outgrowths and tectonic distortion. Therefore, it is possible to recognize graded bedding at some localities, and this is part of the evidence for the widespread inversion of the Furulund Group (Kirk & Mason 1984). Local variations in rock composition are also reflected in variations in the metamorphic mineral assemblages, one distinctive type showing **garben texture** in amphibole porphyroblasts (Fig. 2.3).

The progressive increase in metamorphic grade from southeast to northwest is marked by a general increase in the grain size of the metamorphic minerals, although this may be masked by local variations. The phyllosilicate minerals tend to show a steady increase in grain size, from phyllite to schist (Vogt 1927). Garnet and amphibole occur as porphyroblasts, which tend to become larger as the metamorphic grade increases.

The clearest indication of the increase in metamorphic grade is the incoming of garnet porphyroblasts, which are absent in the low-grade parts of the area, but present in the high-grade parts. Their appearance marks a change in the mineral assemblages in the more iron- and aluminium-rich layers of the Furulund schist. Figure 7.9 shows the line on the map where the garnets first appear, and which can be located on the ground to within approximately 30 m. The rocks north and west of this line lie in a garnet zone of regional metamorphism (or in a higher zone), the type of metamorphism being similar to the Barrovian metamorphic zones of the Grampian Highlands (Vogt 1927, Turner 1981). In the left centre of Figure 7.9, two bands of schist are intercalated into the Sulitjelma amphibolites. When the southern strip is followed on the ground from east to west across the small lake Otervatn, the schist is found to be free of garnets on the eastern side, while on the western side small porphyroblasts of almandine-rich garnet can be seen. Thus, in the schist of comparatively uniform composition in the strip, almandine garnet is absent from the mineral assemblage on the eastern side of the lake, but present on the western side. The conditions of metamorphism must have changed in some way, i.e. the metamorphic grade increased, by a critical amount, so that garnet formed on one side, but not on the other. A line, known as an isograd or line of equal metamorphic grade, crosses the strip of schist at this point (Fig. 7.9), and marks the low-grade limit of the garnet zone. It is therefore called the 'garnet isograd'. Although the garnets are small, they can easily be recognized in the field with the aid of a hand lens, and hence the isograd can be mapped directly. This is the way most isograds have been mapped (including those in Scotland), but there are a few examples which have been mapped by the study of many thin sections.

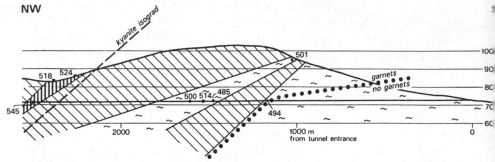

Figure 7.10 Cross section along the hydroelectric tunnel, showing the inverted garnet isograd, with garnet-bearing rocks above, and garnet-free rocks below. Numbers are calculated peak temperatures in °C for rocks from the tunnel and the surface.

In the Sulitjelma example, the isograd mapped in the field was subsequently confirmed by study of thin sections (Vogt 1927). The incoming of garnet is not only seen at the surface. A hydroelectric water supply tunnel cuts through the Sulitjelma amphibolites and the Furulund schist a little to the west of Otervatn (Fig. 7.9). Figure 7.10 shows a cross section along the line of the tunnel, and illustrates the fact that the isograd is not strictly a *line* but a *surface*, and its local direction may be expressed as a dip and strike.

Isograds are so important in the discussion of regional metamorphic terrains that it is worth pausing to define them further (Tilley 1924). Like the garnet isograd at Sulitjelma, an isograd surface may be defined by the *first* appearance of a new mineral with increasing metamorphic grade in a regional metamorphic terrain. If we wish to be precise in naming the isograd, it should be called a 'garnet-in isograd', because garnet comes into the mineral assemblages with increasing metamorphic grade. It need not appear in all horizons of any particular stratigraphical unit, but only in those of a suitable composition. For example, porphyroblasts of garnet only appear in the more iron- and aluminium-rich layers of the Furulund schist. But layers of suitable composition must obviously be fairly common for an isograd line to be mapped. In the case of the garnet isograd at Sulitjelma, the variation in rock composition occurs on a scale of 1 cm to 1 m, so most outcrops contain some layers of suitable composition.

An isograd can be defined in other ways than the incoming of a mineral with increasing metamorphic grade. It may be defined by the *disappearance* with increasing grade of a mineral which is stable at low grade but not at high grade. For example, chlorite is absent from the highest grade schists at Sulitjelma, so theoretically, a 'chlorite-out isograd' could be mapped by the disappearance of chlorite. However, because chlorite does not form porphyroblasts like garnet, and because it is often found in the garnet and higher zones, formed after the peak of metamorphism on the retrograde part of the P–T–t path, such an isograd has not been mapped at Sulitjelma.

It is also possible to define isograds by the incoming or disappearance of *pairs* of minerals with increasing grade, but again this has not proved practicable at Sulitjelma.

Figure 7.11 is a drawing of a thin section of a schist from below the garnet isograd. Bedding is picked out by a layer which is richer in quartz and plagioclase feldspar crystals (b–b). These crystals have irregular outlines, not parallel to special crystallographic directions, and are clastic grains which survive from the original sedimentary fabric (Kirk & Mason 1984). By contrast, the phyllosilicate minerals have recrystallized from original very fine-grained clay minerals and volcanically derived chlorite, to much coarser-grained metamorphic muscovite, biotite and chlorite. These phyllosilicate minerals define a schistosity, i.e. they show a strong tendency to lie in one direction. They are plate-like in their crystal habits, and therefore the schistosity is a planar preferred orientation or foliation. This preferred orientation is only a statistical predominance of one preferred direction and individual phyllosilicate flakes may be oblique to the schistosity plane or even at right angles to it, although the number of such flakes is less than of those which are parallel or nearly parallel to the schistosity direction. It is obvious that the schistosity plane (s–s) is oblique to the bedding plane (b–b). The preferred orientation was imposed during deformation of the Furulund schist into folds. On the outcrop scale it can be seen to be an axial planar schistosity to a set of folds.

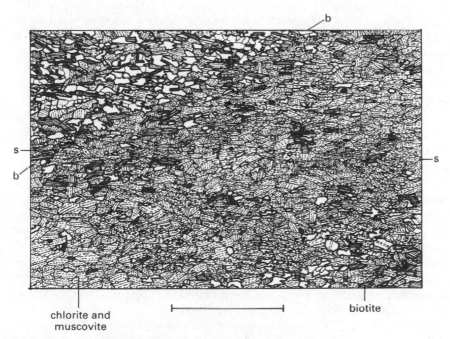

Figure 7.11 Thin section drawing of schist from below the garnet isograd, at the western end of Lomivann. s–s schistosity direction, b–b bedding. Scale bar 1 mm.

Closer inspection of the schistosity reveals a more complicated pattern, especially in the phyllosilicate-rich layer alongside the quartz and feldspar-rich bed. A segregation of minerals can be recognized, with layers particularly rich in phyllosilicates, alternating with layers containing quartz and feldspar. This is a compositional banding induced during metamorphism by the process of pressure solution. Quartz and feldspar dissolved away along the mica-rich layers. Finally, in the quartz- and feldspar-bearing bands between the mica-rich layers, a different preferred orientation is preserved in some micas, making a high angle to the main schistosity. This is an earlier schistosity (s_1), mostly overprinted, but locally surviving.

The history of the textural evolution of this rock may therefore be summarized as follows:

(1)   Deposition as a finely laminated turbidite
(2)   Imposition of an early schistosity (s_1), only locally preserved
(3)   Folding associated with an axial planar schistosity (s_2) oblique to the bedding, and with pressure solution producing a compositional banding into mica-rich with quartz and feldspar rich layers

Figure 7.12 shows a rock of approximately the same metamorphic grade as that in Figure 7.11, in which the two different schistosities are more distinct. The family of parallel surfaces (s_2) shows a different, local orientation of the muscovite, chlorite and biotite flakes from the penetrative schistosity (s_1). These surfaces define a crenulation cleavage, which is a non-penetrative fabric because it is only developed by part of the population of phyllosilicate flakes in the rock. It can be seen that the muscovite and chlorite flakes defining the s_1 schistosity change their orientation into the new direction parallel to the s_2 crenulation cleavage zones as the zones are approached. This sequence of events may be confirmed by field study of the geological structures, which shows that specimens of Furulund schist showing crenulation cleavage are confined to the hinge-zones of a set of folds which fold the s_1 schistosity (Nicholson 1966).

It can also be seen in Figure 7.12 that larger biotite porphyroblasts show a preferred orientation parallel to the crenulation cleavage s_2. This indicates that the porphyroblasts grew while the s_2 cleavage was developing, and the associated folds forming (or perhaps later). The few biotite flakes in Figure 7.11 also do not follow the penetrative schistosity direction as closely as the chlorite and muscovite flakes, suggesting that they too grew later than the imposition of the s_1 schistosity. This observation on both thin sections indicates that the s_1 schistosity evolved before the metamorphic peak temperatures were reached and biotite crystallized.

Figure 7.13 shows a thin section of Furulund schist from above the garnet-in isograd. The metamorphic assemblage has now changed. The

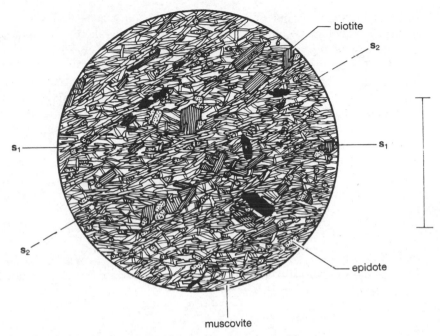

Figure 7.12   Biotite schist from below the garnet isograd, north of Lomivann. s_1–s_1 penetrative early schistosity, s_2–s_2 later spaced cleavage. Scale bar 1 mm.

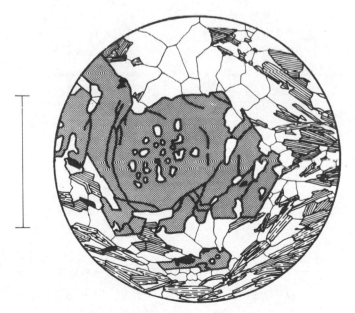

Figure 7.13   Garnet–mica schist, Otervatn. Scale bar 1 mm.

rocks in Figures 7.11 and 7.12 have the assemblage quartz + plagioclase + muscovite + chlorite + opaques + epidote + biotite, while that in Figure 6.5 has the assemblage quartz + plagioclase + muscovite + chlorite + opaques + epidote + biotite + almandine-rich garnet, although all three rocks have approximately the same compositions. The first two AFM diagrams in Figure 7.14 show the change in mineral assemblages, the incoming of garnet, as a consequence of the increase in metamorphic grade. As discussed earier, the incoming of garnet is due to a metamorphic reaction, which has occurred in the rocks above the garnet-in isograd, but not in the rocks below. The close similarity in chemical composition between garnet and chlorite, and the disappearance of chlorite with increasing metamorphic grade, both imply that chlorite breaks down to help form garnet. But the reaction cannot be simply:

$$chlorite = garnet + H_2O$$

because the Fe: Mg ratios of garnet and chlorite are markedly different, as

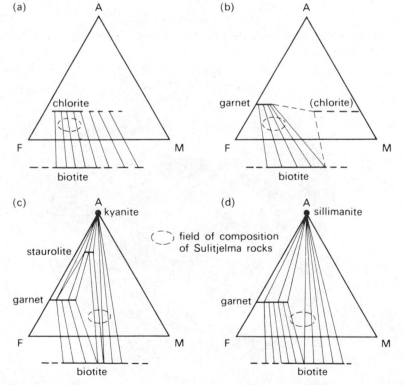

Figure 7.14 AFM triangles showing the change in mineral assemblages of pelitic rocks at Sulitjelma. (a) Biotite zone. (b) Garnet zone. (c) Kyanite zone. (d) Sillimanite zone.

the AFM triangles in Figure 7.14 show, and chlorite persists at Sulitjelma above the garnet isograd, as a prograde mineral, as well as a retrograde mineral. A more plausible reaction involves a change with increasing grade in the Fe^{+2}: Mg ratios of chlorite and garnet which are in thermodynamic equilibrium together at any particular temperature. The reaction could be written

$$\text{Fe-rich chlorite} = \text{Mg-rich chlorite} + \text{garnet} + H_2O.$$

This is a sliding reaction, i.e. the temperature of equilibrium at the surface of the isograd (close to the peak metamorphic temperature) depends upon the Fe^{+2}: Mg ratio of the rock. There is only a recognizable isograd surface, independent of the direction of the stratigraphical boundaries, because this ratio falls in a narrow range of values throughout the Furulund schist.

The relationship between isograds and metamorphic reactions makes isograds of great significance in determining the conditions of metamorphism. As explained in Chapter 2, metamorphic reactions may be investigated experimentally, and the P–T conditions associated with them can also be obtained by thermodynamic calculation. If we assume that there is relatively little overstep of metamorphic reaction boundaries under natural conditions, an isograd surface indicates a particular set of P–T conditions within the volume of rock at peak metamorphic temperatures, which can be discovered by laboratory and calculation methods. This is the fundamental approach of most modern studies of regional metamorphism, which are yielding extensive arrays of temperature and pressure values. Mapping isograds for known metamorphic reactions make it possible to extrapolate P–T values, usually obtained from samples collected at the ground surface, into three-dimensional patterns in the metamorphic rock volume. These patterns may then be compared with those predicted by thermotectonic modelling.

The garnet porphyroblasts in Figure 7.13 contain inclusions of minerals incorporated from the surrounding finer-grained groundmass during the growth of the garnet. They are arranged in definite zones, which are matched by a variation in the chemical composition of the garnet crystals (Burton et al. 1989). In the centre of the crystal the inclusions are small (zone z_1), then there is a zone free of inclusions (z_2), and finally near the outside there are larger inclusions, some of them connected with the granoblastic groundmass surrounding the garnet (zone z_3). The outer inclusions show that the garnet crystals have developed their poikiloblastic habit by growing faster along grain boundaries in the surrounding matrix, and replacing the interiors of the quartz and feldspar crystals more slowly. Although the garnet crystals are poikiloblastic, their grain boundaries against the groundmass quartz and feldspar tend to run parallel to the side of a hexagon (as viewed in the thin section), which represents a cross section through the rhombic dodecahedron

a)

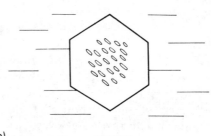

b)

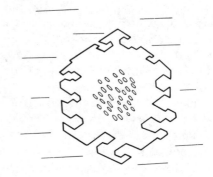

c)

Figure 7.15 Progressive growth of garnet porphyroblasts in the Furulund schist. (a) Nucleation of garnet in fine-grained schist. (b) Growth and rotation of garnet, as nappes advance, and the stratigraphical sequence is overturned. (c) Outgrowth of garnet over coarse-grained groundmass, near peak metamorphic temperature. (d) Deformation of fabric round garnet porphyroblast, during cooling and uplift.

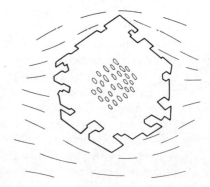

d)

(face indices $\{110\}$), a very common crystal habit in the garnet group of minerals. The tendency to develop faces with low numbered crystallographic indices is an indication that the surface energy of the garnet was very high when it was growing.

It is clear from the distances between the inclusions that the grains enclosed in the cores of the garnets were part of a fabric which was finer grained than the present groundmass. The tendency for the inclusions in zone z_1 to trace out a planar preferred orientation shows that there was already a schistosity (s_1) in the schist when the garnet began to grow. The s_1 schistosity plane now makes a large angle with the schistosity plane outside the garnet crystals (s_0). During the period of growth of the garnets, the internal schistosity plane was rotated relative to the external schistosity plane by deformation (Wilson 1972). Figure 7.15 shows the stages in the evolution of the fabric of the schist, relative to the growth of the garnet crystals.

Thus the porphyroblasts in the Furulund schists at Sulitjelma reveal a complex history of growth of the mineral during the deformation of the schist. The three zones z_1, z_2 and z_3 represent successive stages in the growth of the garnet, and the coarsening in grain size between z_1 and z_3 shows that the temperature was increasing. The small size and close spacing of the inclusions in z_1 indicate that this zone grew before the matrix had become as coarse-grained as it is now, probably when the temperature was lower. The growth of the garnet porphyroblasts preserves part of the history of the evolution of the schist sample along its P–T–t track.

The story can be taken further. The garnet in Figure 7.13 has been analysed at many points, using the electron microprobe (Potts 1987). This reveals that the garnet in the textural zones z_1, z_2 and z_3 has different chemical compositions. It does not change sharply at the boundaries, but shows a gradual change from the centre to the edge. Figure 7.16 shows the profile of chemical composition along a transect through the garnet.

At any particular time (t) during the growth of the garnet, the composition of the garnet depended on the composition of the other minerals with which it was in equilibrium, and on the temperature and pressure. At the temperatures at which these garnets grew, Fe, Mg, Al, Mn and Ca travelled so slowly through the garnet by diffusion that as each layer grew, the older layers inside it were unable to react with the matrix minerals. By contrast, diffusion in biotite and chlorite was rapid. This has been confirmed by electron microprobe analysis, which shows that while the garnets are zoned, biotite, muscovite and chlorite crystals are unzoned. For the last stage of growth of the garnet, the temperature and pressure of metamorphism has been calculated from the composition of the rims of the garnets (which are uniform) and the compositions of adjacent crystals of biotite and plagioclase feldspar. This is another example of geothermobarometry. For the zoned garnets, the calculation may be taken further (Spear et al, 1984) It is assumed that the

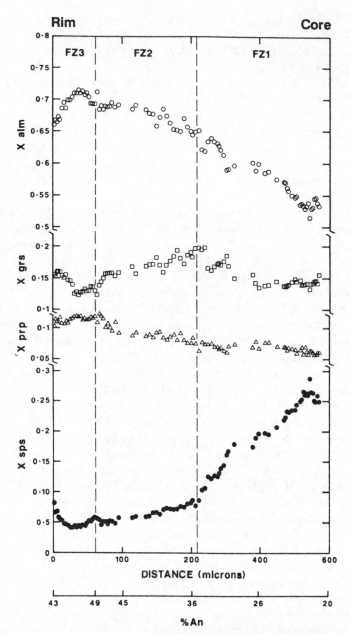

Figure 7.16 Variation of composition of a garnet porphyroblast from rim to core, expressed as proportions of end members spessartine $Mn_3Al_2Si_3O_{12}$ (sps), pyrope $Mg_3Al_2Si_3O_{12}$ (prp), grossular $Ca_3Al_2Si_3O_{12}$ (grs) and almandine $Fe_3Al_2Si_3O_{12}$ (alm). From Burton *et al.* (1989).

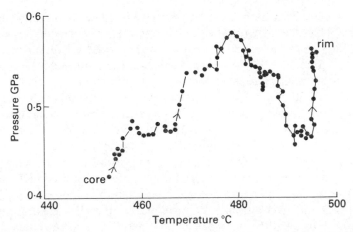

Figure 7.17 Part of a P–T–t path derived by calculation from the zoned garnet porphyroblast in Figure 7.16.

total amounts of Fe, Mg, Ca, Al and Mn in the rock remained unchanged throughout the time when the garnet porphyroblasts grew. The amount of these elements trapped in the inner zones of the garnet may be calculated, and so the remainder available to crystallize in biotite, muscovite and plagioclase feldspar can be calculated. From the observed compositions of the inner zones of the garnet, and the calculated compositions of the co-existing garnet, muscovite and feldspar, a continuous series of temperatures and pressures experienced by the garnet porphyroblast and its matrix can be calculated, using a computer program. Part of the P–T–t track for the rock calculated in this way is shown in Figure 7.17 (Burton *et al.* 1989).

This conclusion has been reached by making a long chain of assumptions:

(1) that the P–T conditions of the garnet-forming reaction have been correctly calculated from experiments and thermodynamic calculations.
(2) that there was little overstep of the reaction temperatures during garnet growth.
(3) that there was no diffusion of elements through the garnet porphyroblasts, but very quick diffusion through biotite, plagioclase and quartz, and very quick exchange of elements between the outer layer of the garnet and the matrix.
(4) that there was only one P–T–t cycle.

Metamorphism of the Sulitjelma amphibolites

The Sulitjelma amphibolites represent the upper layers of the Sulitjelma ophiolite, including the pillow lavas, sheeted dyke complex and part of the gabbro layer. Many primary igneous features survive, including pyroclastic fragments of basic composition, flattened by tectonic deformation but still recognizable, pillow structures with their fine-grained chilled margins, sheets with chilled margins at their contacts and primary columnar jointing (Fig. 7.18). Smaller scale structures visible in the field include porphyritic texture, with plagioclase phenocrysts most common but also pyroxene phenocrysts, both replaced by pseudomorphs of metamorphic minerals. Where the amphibolites include coarser-grained gabbros, ophitic texture is preserved, again by pseudomorphs after pyroxene and plagioclase feldspar. The lavas, sheets and gabbros have many of the geochemical characteristics of mid-ocean ridge basalt (**MORB**), but there are also features which suggest formation in a back-arc basin rather than a large ocean basin (Boyle 1987, 1989). The metamorphic assemblages of the amphibolites were all formed during the Scandian collision, none surviving from earlier ocean-floor metamorphism. However, the nature of the copper ores, and the chemistry and petrology of the metabasites associated with the ores, indicate that suboceanic hydrothermal systems did exist, and these presumably caused metamorphism of ocean floor type. As the name amphibolite implies, the Sulitjelma amphibolites contain essential calcic amphibole (hornblende or actinolite), which can be recognized in hand specimen in the coarser-grained specimens. The amphibolites cross the garnet isograd, so that it is possible to study metamorphic assemblages from either side of the isograd surface.

Figure 7.18 Sheeted dykes in the Sulitjelma Ophiolite, showing preserved columnar jointing and chilled margins.

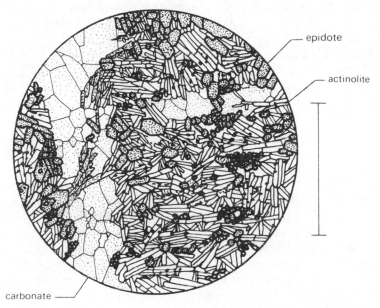

Figure 7.19 Thin section from a pillow lava in the Sulitjelma Ophiolite, from below the garnet isograd. Scale bar 1 mm.

These will be illustrated by considering two rocks, one from a relatively low-grade part of the area, below the garnet isograd, the other from a high-grade part.

The isograd surfaces in the amphibolites cannot be recognized in the field, although there is a distinct change from green rocks in the low grade parts of the area to green-black rocks in the high-grade parts. As in the Furulund schists, there is a tendency for the grain size to increase with metamorphic grade. However, because primary igneous textures and structures are widely preserved, there is more local variation in grain size than in the schists.

Figure 7.19 shows a thin section of a massive, pale-green amphibolite from inverted pillow lava in the Swedish part of Sulitjelma. The rock has the mineral assemblage actinolite + epidote + carbonate + plagioclase + quartz and is cut by veins about 1 mm thick composed of carbonate, clinozoisite and chlorite, although chlorite is not an essential mineral in the rock itself.

The amphibole is colourless and occurs as clusters of radiating prismatic crystals. The prismatic habit of the amphibole enables the extinction angle to be determined, and it is 20°, a value characteristic of actinolite rather than of hornblende. The composition is confirmed by microprobe analysis. The distinction between actinolite and hornblende is not always easy to make by petrological microscope, when amphibole crystals are small and may not have well-developed prismatic habit, but it is always worth making an attempt on metamorphosed basic igneous rocks. There is no preferred radiating habit makes the rock unusually tough and resistant to breaking with a geological hammer. Interstitial to the amphibole crystals are crystals of plagioclase. They may be identified because one or two crystals show wedge-shaped lamellar twins due to tectonic strain, and flat cleavage planes may be seen in others. The extinction angle is less than 15° and the grains have refractive indices close to that of the mounting medium (1.53). This suggests that the plagioclase is albite (An_{1-10}), a conclusion which is supported by microprobe analysis. The quartz can be distinguished by slightly higher refractive indices, birefringence and uniaxial positive interference figures.

The other major mineral of this rock is epidote. It occurs as stumpy prismatic crystals with rounded outlines and a large range of grain sizes. It has appreciably higher positive relief than the amphibole, and between crossed polars displays the anomalous interference colours characteristic of all minerals of the epidote family. The colours go up to second order green, indicating that the mineral in the assemblage of the main part of the rock is epidote, contrasting with clinozoisite (which shows low first-order interference colours) occurring in veins with calcite. The rock also contains accessory sphene which has even higher R.I.s than the epidote, a brown body colour, and high birefringence.

The rock shown in Figure 7.20 comes from a higher-grade part of the

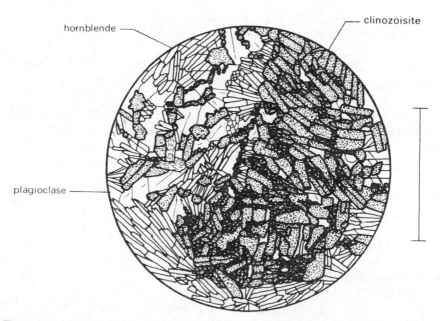

Figure 7.20 Amphibolite from the Sheeted Dyke Complex, above the garnet isograd, showing a pseudomorph of clinozoisite and plagioclase after a plagioclase phenocryst in porphyritic dolerite. Scale bar 1 mm.

Sulitjelma amphibolites, in the sheeted dyke complex. It is darker green in hand specimen than the previous rock, because the amphibole is hornblende, not actinolite. Its most conspicuous feature is the presence of white patches, which are pseudomorphs after plagioclase phenocrysts in the dolerite of the parent dyke, now consisting of crystals of clinozoisite and plagioclase feldspar. The clinozoisite has a prismatic habit and a rather variable grain size. The surrounding part of the amphibolite contains the mineral assemblage hornblende + plagioclase + epidote + sphene + opaques. The maximum extinction angle of 27° indicates that the amphibole in the groundmass is hornblende, not actinolite. The epidote mineral is identified as epidote by its second-order anomalous interference colours and biaxial negative conoscopic interference figure. When the rock is studied under low magnification, it can be seen that the hornblende crystals occur in parallel or radiating clusters, with patches of epidote and plagioclase in between. They probably represent pseudomorphs after the pyroxene and plagioclase, respectively, in the parent igneous rock.

The presence of two different members of the epidote group shows that two different metamorphic assemblages crystallized in the rock, one in the pseudomorphs, one in the groundmass, i.e. there are two contrasting **domains of equilibrium**. The assemblages differ in the chemical potentials of aluminium and ferric iron, the chemical potential energy of Al being higher in the pseudomorphs, the chemical potential of Fe^{+3} higher in the groundmass. The higher chemical potential of Fe^{+3} in the groundmass caused the crystallization of Fe-rich epidote, the lower chemical potential of Fe^{+3} in the pseudomorphs caused the crystallization of Fe-poor clinozoisite. At the time of metamorphism there was a chemical potential gradient between the groundmass and the pseudomorphs, and because Fe^{+3} and Al are relatively insoluble, these elements could only travel over small distances (considerably less than the 1–2 mm of the pseudomorphs) during the whole time of metamorphism. The existence of a local gradient in activity of Fe^{+3} is indicated by clinozoisite grains near the edges of the pseudomorphs, which have a slightly higher birefringence than those in the cores, implying that a small amount of Fe diffused into the pseudomorphs during the whole time of metamorphism. This rock contains two domains of equilibrium, with different chemical compositions – the Ca and Al-rich domain within the pseudomorph after plagioclase feldspar, and the more Mg- and Fe-rich domain representing the former groundmass. The plagioclase of both pseudomorphs and groundmass shows quite abundant lamellar twinning and the composition may be determined as An_{28} in both cases.

When the mineral assemblages of the groundmass of this rock is compared with the rock of Figure 7.19 two differences can be recognized which are caused by the difference in metamorphic grade, not the small difference in chemical composition. Hornblende has taken the place of actinolitic amphibole, and calcic oligoclase has replaced albite as the plagioclase mineral.

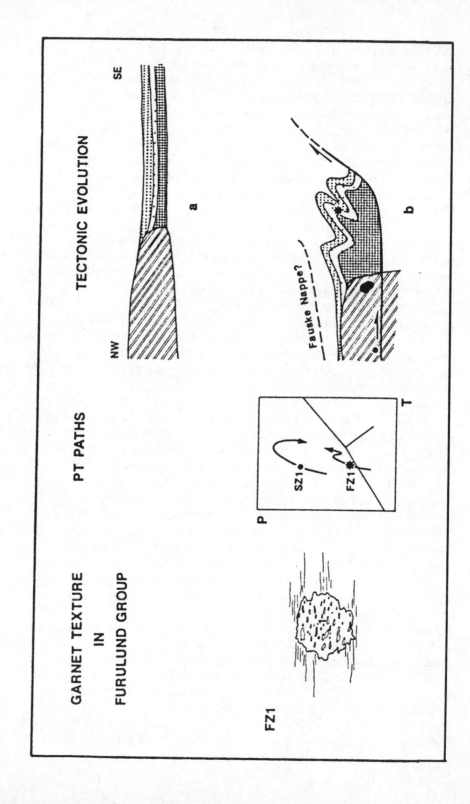

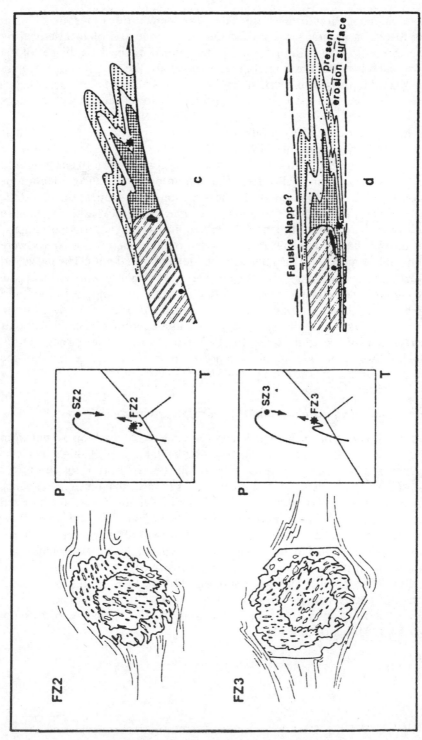

Figure 7.21 Garnet growth stages, P–T–t paths and tectonic evolution diagrams from the Sulitjelma region, from Burton *et al.* (1989). Compare Figure 7.15.

These changes reflect the increase in the peak temperature of metamorphism, and there is a general tendency for the hornblende and oligoclase-bearing assemblages to be within the garnet zone, defined by the assemblages of the Furulund schist. But the amphibolites do not show the sharp changes seen at the isograd in the schists, particularly because hornblende and calcic plagioclase are found in amphibolites at considerably lower metamorphic grades than the isograd.

The basic igneous rocks at Sulitjelma are slightly magnetic. Study of their directions of magnetization (Piper 1974) have interesting implications for the metamorphism. The method of study is that cores about 10 cm long and 2.5 cm in diameter are drilled from the rock in place, and their orientations recorded before they are removed. The direction of magnetization within the rock is then measured in the laboratory. The basic rocks did not acquire their present magnetic direction during cooling from the original magma, but during the metamorphism when the original igneous pyroxenes were replaced by amphiboles (usually hornblende) and iron oxides. The preserved direction of **remanent magnetization** of the rocks may be compared with the direction of the magnetic field in Europe at different geological times, and fits best with the direction of the field in Silurian to Devonian time, during the final erosion and uplift stage of the Scandian continental collision. This correlates with the evidence of the P–T–t paths based on garnet zoning, which showed that the peak of metamorphism came towards the end of the deformation history, during a period of erosion and uplift.

The petrological evidence and the thermotectonic model

The P–T–t path from Sulitjelma shows the same clockwise loop as that from Connemara, but in the Norwegian case the later low P, high T metamorphism cannot be correlated with an episode of heating of the upper lithosphere by the intrusion of large amounts of basic magma. The sequence of events is summarized in cartoon form in Figure 7.21. The inverted metamorphic gradient was imposed during the period of tectonic overthrusting, and preserved during the final stage of growth of the garnet porphyroblasts. This implies that uplift must have been relatively rapid, causing a rapid drop in pressure, accompanied by a relatively small increase in temperature. This proposed history is supported by the P–T–t paths constructed from the garnet zoning profiles. Thus the petrological evidence can be explained using a relatively simple tectonic model, which resembles the overthrust Model II quite closely.

8 Metamorphism in stable continental crust

Most of the continental crust in the stable interiors of the continents consists of crystalline rocks with approximately granitic composition, partly or completely overlain by unmetamorphosed sedimentary rocks. The North American continent is a good example, with the crystalline **basement** exposed in the northern part, where it is known as the Canadian Shield, and in a number of inliers in the interior of the continent (Fig. 2.2). Study of the petrology, geological structure and geochemistry of ancient shields shows that the majority of rocks in them were formed by extensional and collisional plate movements, along with volcanic activity, during Proterozoic and later Archaean times (Windley 1984). It has been suggested that other tectonic mechanisms may have occurred in early Archaean times, and some of the evidence used to support this point of view comes from metamorphic rocks. This point will be discussed again later in this Chapter.

In north west Colorado, part of the continental crust of North America has been uplifted, and the Colorado River has excavated a deep gorge, the Grand Canyon, through Mesozoic and Palaeozoic sedimentary rocks to Precambrian rocks beneath. The Precambrian rocks are of two types, older schists and granitoids, overlain by Precambrian sedimentary and volcanic rocks, which have been tilted and faulted, but have not undergone meta-

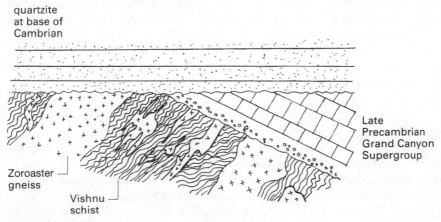

Figure 8.1 Relationships of sedimentary and metamorphic rocks at the bottom of the Grand Canyon, Arizona, USA.

morphism. The age of the metamorphic rocks has been determined by radiometric dating to be 1700 Ma, showing them to belong to the Proterozoic period.

In the Proterozoic metamorphic complex at the bottom of the Grand Canyon, it can be seen that the granitoids are younger than the schists. In many places the granitoids form veins, which were intruded among the surrounding schists (Fig. 8.1). The veins formed when the temperature of the rocks was high, at the same time as the metamorphism of the schists. They have the mineral and chemical composition of granitoids, so that much of the metamorphic complex can be described as a mixture of granitoid veins with the metamorphic schists. The rest of the complex consists either of schist, or of larger bodies of granitoids. Where granitoids and schists are too intimately mixed to be shown separately on geological maps, the rocks of the complex are described as migmatites. The name means 'mixed rocks', the mixtures concerned being between granitoid rocks and metamorphic rocks (usually metamorphosed sediments).

There has been controversy about the mode of origin of migmatites in the past, but a combination of petrological and geochemical study of migmatites with consideration of the results of experimental petrology has largely resolved the arguments (Read 1957, Ashworth 1985). Migmatites formed by partial melting of rocks during metamorphism. The granitoid melts produced by the partial melting may remain close to the place where they formed, or they may migrate shorter or greater distances, to give rise to the intrusive field relationships which are shown in Figure 8.1. Figure 8.2 shows a set of granitoid veins, surrounded by a darker fringe of biotite-rich schist, which gives way outwards to rather lighter coloured schist. The granitoid veins are referred to as the **leucosome** of the migmatite, the dark fringing part of the schist as the **melanosome** and the rest of the schist as the **mesosome** (Fig. 8.3). The way in which the melanosome fringes the leucosome veins shows that their origins are linked together in some way. One possible interpretation is that the granitoid veins originated by partial melting of the schist immediately surrounding them. In this case the melanosome may have arisen because the quartz and feldspar minerals in the schist would have been incorporated into the melt in higher proportions than the biotite, which was therefore concentrated in the edges of the unmelted schist. This interpretation accords well with what is known of melting relationships in systems approximating to granitoid compositions (Cox *et al.* 1979).

This interpretation fails when we consider the relative volumes of leucosome, mesosome and melanosome in Figure 8.3. If we assume that the mesosome represents the schist protolith, the melanosome schist minus partial melt of leucosome composition, and the leucosome veins extracted partial melt, a glance at the figure shows that there is too much leucosome in the veins to have been extracted from the relatively narrow melanosome fringes. A comparable problem is encountered when migmatites are analysed

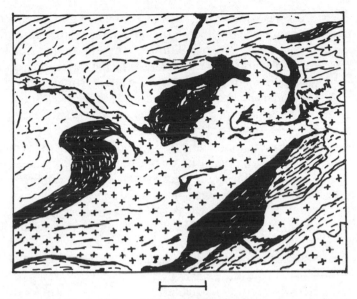

Figure 8.2 Granite veins in high-grade Vishnu schist, Grand Canyon. Scale bar 10 cm.

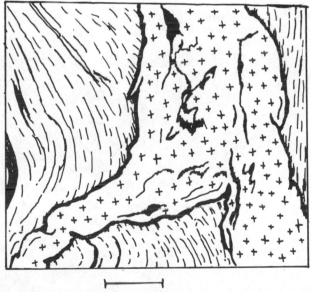

Figure 8.3 Granitic vein partly cross-cutting, partly following schistosity of Vishnu schists. It is surrounded by a melanosome of mica-rich schist. Scale bar 10 cm.

chemically, because it is not usually possible to combine the compositions of leucosome and melanosome together in the observed proportions of the two rock types and arrive at the composition of the mesosome. Some other process must be involved as well. In the case of the migmatites shown in Figure 8.3, it may be that part of the leucosome has crystallized from partial melt derived from the melanosome, but that most of the leucosome formed by partial melting elsewhere (perhaps not very far away), migrated through the melting schist, and was intruded in its present position.

The migmatites of the Grand Canyon do not display gneissose banding. True banded gneisses with migmatite composition are very widespread in other Precambrian shield areas, and Figure 8.4 shows an example from Wadi Bisha, in the Arabian Precambrian Shield, Saudi Arabia. This rock has its leucosome and mesosome arranged as alternating layers, 1 mm to 5 mm thick, composed of granitoid and mica-rich schist. Such migmatites may be displaying the effects of partial melting *in situ*, the leucosome layers representing crystallized partial melt, the mesosome layers the residue of unmelted schist. If this were the case, the partial melt would have migrated only a few millimetres from its site of origin. Migmatites of this type are often described as **lit-par-lit** migmatites, and are probably the commonest type of gneiss.

Figure 8.4 Lit–par–lit   migmatite, with interlayering of granitic rocks (light colour) and schist (darker), the veins of granite being approximately parallel to the schistosity. Wadi Bisha, southern Saudi Arabia.

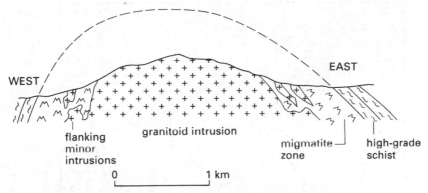

Figure 8.5 Zone of migmatites surrounding a granitoid intrusion in the core of an anticline, Jebel Dhuhya, southern Saudi Arabia.

The migmatites seen in this part of Wadi Bisha are associated with a body of granodiorite forming the mountain Jebal Dhuhya (Fig. 8.5). The granodiorite occupies the core of an antiform, and in the surrounding region the metamorphic grade increases towards the antiform (Gass 1979). Although the migmatites are found over a large area, the zone over which they occur is relatively thin, and they are folded up into a gentler arch than the Jebal Dhuhya Antiform itself. This geometrical relationship may be explained by the mechanism of introduction of heat into this part of the Arabian Shield in Proterozoic times. The heat was supplied by the intrusion of large amounts of magma of intermediate to acid composition, of which the granodiorite is a sample. The country rocks were basic igneous rocks and greywackes, collectively denser than the granodiorite magma, which therefore moved upwards as a diapir, arching the surrounding metasediments, but cooling itself as it did so. The result was a pattern of isotherms which are less folded than the original bedding of the country rocks. Migmatites began to form when the rocks reached their melting temperatures. The melting temperature of the greywackes would have been lower than that of the amphibolites, because the composition of the greywackes is more suitable for melting, and the migmatites are more widely developed in them as a result.

Since migmatites apparently represent either metamorphic rocks partially melted *in situ*, or rocks where a granitoid partial melt has migrated only a short distance into high-grade metamorphic rocks, close to the time of metamorphism, it is clear that many of the rocks of Precambrian shields went to high temperatures at the peak of metamorphism. The onset of partial melting would provide an upper temperature limit to metamorphism, because if heat continued to be supplied to the rocks, instead of a further increase in temperature, a large proportion of the rock would melt. In many areas of metamorphic rocks, in more recent orogenic belts as well as in the ancient shields, this does appear to be the upper temperature limit of

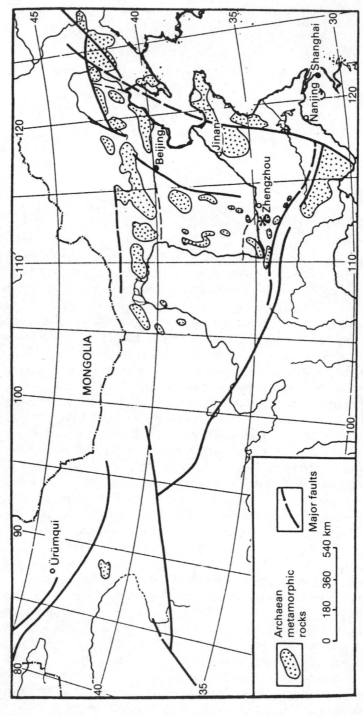

Figure 8.6 Areas of Archaean gneisses in China. Asterisk – Dengfeng district, Central Mountains (Song Shan).

metamorphism. This is clearly seen in the progressive regional metamorphic sequence of Connemara, Ireland (Chapter 7).

It must be emphasized finally that high-grade and partially melted metamorphic rocks are not the only rocks covering large areas of Archaean shields. A significant area also consists of metamorphosed mafic to ultramafic igneous rocks, constituting 'greenstone belts' (Windley 1984). These show variable metamorphic grade, broadly comparable with the metamorphic sequence described in Chapter 7 of the metamorphosed basic igneous rocks of Sulitjelma. Although the metamorphism of these rocks is not described in detail in this book, readers should not assume that this type of metamorphism can be neglected in the discussion of the evolution of continental crust.

Metamorphism in Dengfeng County, Henan Province, China

In the Grand Canyon, the metamorphism of the Archean rocks took place before the deposition of overlying unmetamorphosed Proterozoic and Palaeozoic rocks. This is indicated by the presence of an unconformity surface preserving a Proterozoic weathering profile, beneath the oldest Proterozoic rocks. In many continental areas, the history of metamorphism is more complex than this, and the example to be discussed here comes from the stable continental block of northern China (Fig. 8.6). Here Archaean gneisses, schists and migmatites are overlain by the sequence of Proterozoic rocks illustrated in Figure 8.7. The Proterozoic quartzites form abrupt peaks of the Song Shan (Central Mountains) of China, rising up from an undulating plateau of gneisses. Radiometric ages show that the Archaean rocks were metamorphosed and intruded by granitoids 2986 Ma ago (Cheng 1986). The overlying Proterozoic rocks have also been metamorphosed, and underwent several episodes of folding deformation (Ma & Wu 1981, Mason 1989).

The sedimentary succession near the base of the Proterozoic sequence suggests at first that there is an unconformity at the very bottom, as the quartzite passes downwards into conglomerates which include fragments of the Archaean gneisses. The exact contact between Proterozoic and Archaean is frequently buried beneath debris from the overlying cliffs, but at some localities where it is exposed, a layer of phyllite about 10 m thick is found between the lowest conglomerate layer, and the top of the gneiss. Figure 8.8 shows a thin section of this phyllite from a locality called Shi Chuan, close to Dengfeng (Fig. 8.7). Comparison with the phyllonites described in Chapter 4 suggests that this phyllite layer is actually a fine-grained phyllonite, and that the apparent unconformity is marked by a thrust surface between the Proterozoic conglomerate and the Archaean gneiss.

This conclusion is supported by a study of the contact near Shao Lin Dam (Fig. 8.7), where the contact rocks between the Archaean gneisses and

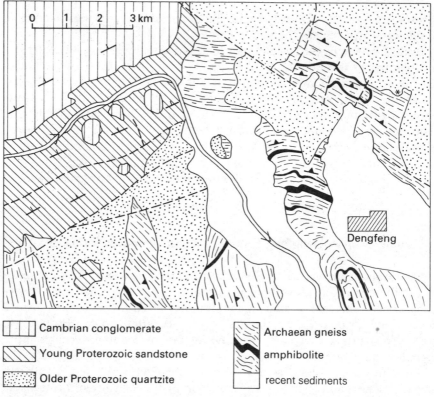

	Cambrian conglomerate			Archaean gneiss
	Young Proterozoic sandstone			amphibolite
	Older Proterozoic quartzite			recent sediments

Figure 8.7 Geological map of the Dengfeng district, Henan Province, China. Asterisk – location of schist shown in Figure 8.8.

Proterozoic quartzite are quartzo–feldspathic rather than schistose. A study of thin sections of the quartzite in this locality, shows that preferred orientation of the quartz grains becomes more and more pronounced as the contact is approached from above (Ma & Wu 1981). In other words, a metamorphosed sandstone changes over a distance of few metres into a quartz-rich mylonite.

The geological map of the Dengfeng area (Fig. 8.7) shows that the Proterozoic rocks have been tightly folded, several times, and comparable folding can be seen on a smaller scale in several places. The early folds show an axial planar cleavage, indicating that the multiple folding was accompanied by metamorphism. During the deformation which caused the folding of the Proterozoic, the Archaean gneisses, which had already been deformed, metamorphosed and intruded by igneous rocks, behaved differently from the thick Proterozoic sedimentary sequence which overlay them. There was a strong tendency for differential movements to be concentrated near the contact surface between the Proterozoic and Archaean rocks, and this has given rise to the phyllonites at Shi Chuan and the mylonites at Shao Lin

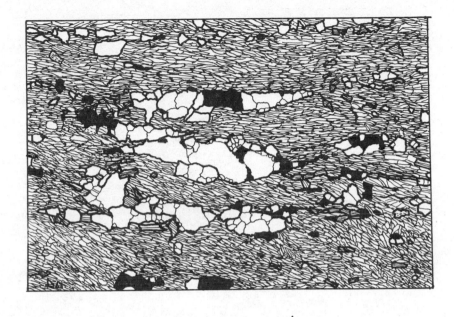

Figure 8.8 Phyllonite from the base of the Proterozoic Quartzite, Shi Chuan, Dengfeng, China. Scale bar 1 mm.

Dam. An unconformity surface not affected by thrusting parallel to the bedding direction has not yet been discovered.

Pyroxene gneisses of northwestern Scotland

The Precambrian gneisses described from Colorado, USA, Saudi Arabia and China show metamorphism of comparable type to that seen in younger collisional zones (Chapter 7), although their metamorphic grade is higher, and melting has occurred. Some areas show even higher temperature metamorphism, to temperatures at which hydrous minerals such as amphiboles and micas begin to break down, and be replaced by anhydrous minerals such as pyroxenes and spinels.

Such rocks are found in the oldest parts of the Lewisian Complex of northwestern Scotland, which is a fragment of the Canadian Shield, separated from the Greenland part by the comparatively recent opening of the North Atlantic Ocean. The predominant rock type is gneiss, often of granitic composition, with well-developed gneissose banding. Radiometric age determinations show that the last metamorphism in the Complex ceased 1700 Ma ago. One group of rocks giving older dates (2900–2200 Ma), the Scourian Complex, exists as relict patches between regions metamorphosed later

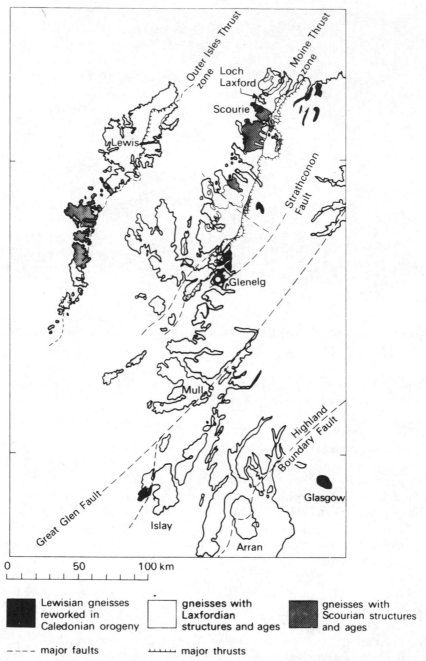

Figure 8.9 Geological map of northwestern Scotland, showing the Lewisian Complex, divided into areas deformed and metamorphosed during the Scourian, Laxfordian and Caledonian orogenic events.

(2200–1700 Ma), the Laxfordian Complex. Rocks of the Lewisian Complex have been remetamorphosed in the northwestern part of the Caledonian orogenic belt (Fig. 8.9). The rocks described in this chapter come from the area around the village of Scourie in the northern part of the mainland area of the Lewisian Complex, where a transition from older rocks of the Scourian Complex to younger rocks of the Laxfordian Complex occurs just north of the village. The rocks are banded gneisses whose composition varies from that of acid igneous rocks to that of ultrabasic igneous rocks. The gneisses are cut by many dykes of basic igneous composition. The banding of the gneisses is caused by variations in rock composition, the darker bands being richer in FeO and MgO, the lighter bands in SiO_2, Na_2O and K_2O. The petrology of the gneisses has been described by Sutton & Watson (1951).

Figure 8.10 shows a gneiss of acid igneous composition. There is banding in the outcrop and in the hand specimen, but the thin section is too small for the variation in mineral composition to be seen. The mineral assemblage is quartz + feldspar + orthopyroxene + clinopyroxene + opaques. The rock also contains hornblende and biotite in reaction rims, which are therefore not members of the high-temperature mineral assemblage listed, but belong to a later, lower temperature stage in the P–T–t cycle. The earlier mineral assemblage has a coarse-grained granoblastic texture. The feldspar is anti-perthite, i.e. crystals of oligoclase (An_{30}) containing plate-shaped inclusions of K-feldspar. Compare it with the perthite in Figure 4.2. The antiperthite

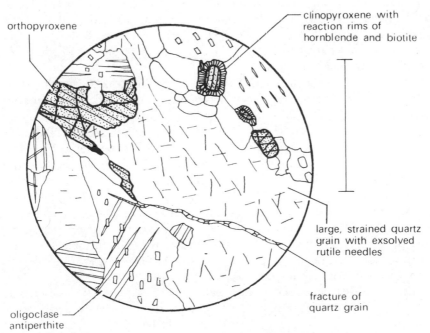

orthopyroxene

clinopyroxene with reaction rims of hornblende and biotite

large, strained quartz grain with exsolved rutile needles

fracture of quartz grain

oligoclase antiperthite

Figure 8.10 Charnockitic gneiss from Scourie, Sutherland, Scotland. Scale bar 1 mm.

has formed because plagioclase feldspar accepts more potassium into its crystal structure at high temperatures than at low temperatures. As the plagioclase cools, the potassium ions become concentrated into volumes of K-feldspar, surrounded by larger volumes of Ca, Na feldspar, so that what was a homogeneous crystal at high temperatures becomes an intergrowth of two mineral types. This process is called **exsolution**. It will only occur if the rate of cooling is low enough for diffusion of K^+ and Na^+ ions through the crystal lattice to have time to make the separation occur. The presence of perthite therefore gives an indication of the rate of cooling on the retrograde part of the P–T–t loop. In the rock in Figure 8.10, the twin lamellae of the plagioclase in the antiperthite are bent, the strain shadows occur in the quartz, showing that deformation occurred in the last stages of the P–T–t cycle, when exsolution was over, and the rock was relatively cool.

This rock has the chemical composition of granodiorite, and the mineral assemblage is similar to the list of minerals in a granodiorite, except that the minor amounts of ferromagnesian minerals are anhydrous orthopyroxene and clinopyroxene, not hydrous micas and amphiboles. The biotite and hornblende in the rock occur in reaction rims which are younger than the main mineral assemblage. The granoblastic texture of the rock in thin section, and the banding seen in hand specimen, show clearly that it is a metamorphic rock, not an igneous one. Rocks with granitic to granodiorite compositions, but with metamorphic textures and anhydrous ferromagnesian minerals, belong to the **charnockite suite**. Charnockite is strictly defined as the member of the suite with predominant K-feldspar (i.e. equivalent to granite in chemical composition). The rock in Figure 8.10, containing plagioclase feldspar rather than K-feldspar, is enderbite, the sodium-rich member of the charnockite suite.

The rocks of the charnockite suite are called after the tombstone of Job Charnock in Calcutta, India, which is made of this rock. The stone was quarried from Madras, in southern India, and brought to Calcutta by sea. Charnockites are very widespread in Precambrian shield areas, including the older parts of the Indian subcontinent, and are characterized by their broadly granitic composition, accompanied by lack of hydrous metamorphic minerals and a dark colour in the outcrop and hand specimen. The dark colour is caused by the exsolution of crystals of rutile (TiO_2) from the quartz in the rocks, and some such needles can be seen in the quartz in Figure 8.10. In the case of the Scourie pyroxene gneisses, the dark colour is not strong, and there has been discussion whether or not they should be described as charnockites, but this book follows Sutton & Watson (1951) in using the term for the broadly granitic gneisses.

There are also mafic rock types among the gneisses, and two examples are shown in Figures 8.11 and 8.12. Both have approximately the chemical composition of basalt. The rock in Figure 8.11 has the mineral assemblage plagioclase (An_{46}) + orthopyroxene + clinopyroxene + opaques + apatite.

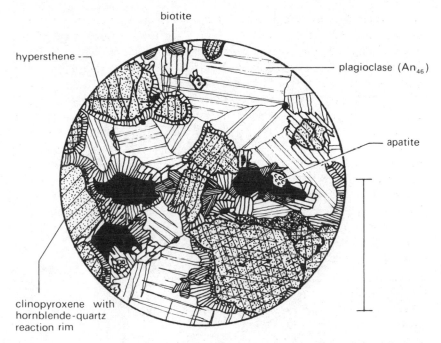

biotite

hypersthene

plagioclase (An$_{46}$)

apatite

clinopyroxene with
hornblende-quartz
reaction rim

Figure 8.11 Pyroxene gneiss of basic igneous composition, Scourie, Sutherland, Scotland.
Scale bar 1 mm.

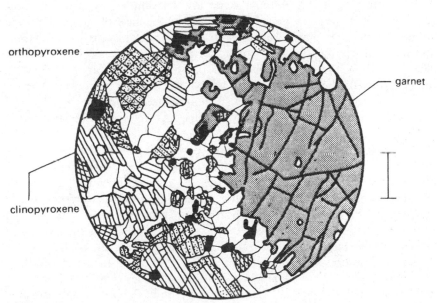

orthopyroxene

garnet

clinopyroxene

Figure 8.12 Pyroxene gneiss of basic igneous composition, with large garnet porphyroblast,
Scourie, Sutherland, Scotland. Scale bar 1 mm.

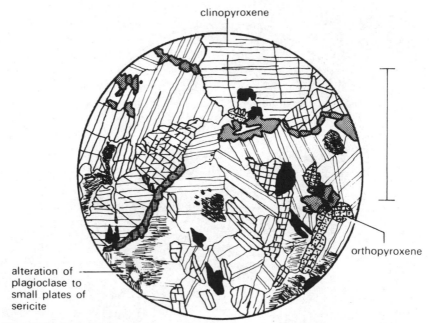

clinopyroxene

orthopyroxene

alteration of
plagioclase to
small plates of
sericite

Figure 8.13 Enlarged portion of the same rock as Figure 8.12, showing interstitial crystals of garnet. Scale bar 1 mm.

The orthopyroxene and clinopyroxene are surrounded by reaction rims, composed of intergrown green hornblende in the case of the rim around clinopyroxene. The plagioclase crystals have lamellar twins showing bent lamellae and strain extinction, similar to the granitic gneiss in Figure 8.10. The minerals in the assemblage list of this rock are similar to those of a tholeiitic basalt, the only difference which can be seen through the petrological microscope being the higher sodium content of the plagioclase. The rock is coarser-grained than basalt, and the texture is granoblastic, rather than micro-ophitic or intersertal (McKenzie *et al.* 1982).

Figures 8.12 and 8.13 show another pyroxene gneiss of basic igneous composition, this time including almandine garnet in the mineral assemblage. There are two textural varieties of garnet: large porphyroblasts (Fig. 8.12) and small crystals forming reaction rims between pyroxene and plagioclase (Fig. 8.13). The porphyroblasts have irregular outlines and contain inclusions of plagioclase, the crystal form reflecting the history of growth of the large garnet crystals. They began by growing as rims to orthopyroxene crystals, gradually replacing them, and then continued to grow along plagioclase–plagioclase grain boundaries. This final stage of growth is clearly shown in Figure 8.13. The core of the porphyroblasts may include garnet which crystallized at the same time as the pyroxenes, making the early mineral assemblage in this rock plagioclase + orthopyroxene +

clinopyroxene + opaques. Compared with the rocks of Figure 8.10, this rock shows more retrograde modification from the later stages in the P–T–t loop. Plagioclase has become clouded by partial alteration to clinozoisite, white mica and sodic plagioclase, showing that hydration occurred at this later stage.

Figure 8.14 shows a gneiss of ultrabasic igneous composition. It has the mineral assemblage clinopyroxene + orthopyroxene + olivine + spinel + - pargasite + opaques. Pargasite is a pale coloured, Al-rich variety of hornblende. The spinel is brownish green in colour. The olivine shows partial alteration to serpentine + magnetite, which occurred in the retrograde stage of metamorphism. It is interesting that in this rock, which has a very Mg- and Fe-rich composition, the hydrous mineral pargasite $(NaCa_2Mg_4Al_3Si_8O_{22}(OH)_2)$ was in equilibrium in the early, high-T mineral assemblage.

The distinctive feature of all the gneisses described from Scourie is the co-existence in the high-T mineral assemblages of orthopyroxene and clino-pyroxene. The presence of either mineral on its own in a metamorphic assemblage does not indicate high-T metamorphism, but the co-existence of the two in equilibrium does. Careful textural study is sometimes needed to distinguish co-existing metamorphic pyroxenes from relict pyroxenes surviving from a parent igneous rock.

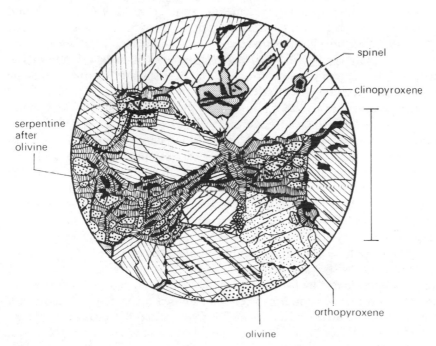

Figure 8.14 Pyroxene–pargasite–olivine gneiss of ultrabasic composition, Scourie, Sutherland, Scotland. Scale bar 1 mm.

The Adirondack Mountains, New York State, USA

The Adirondack Mountains of northern New York are an inlier of Proterozoic rocks of the Canadian Shield, surrounded by Palaeozoic rocks (Bohlen *et al.* 1985). The Precambrian rocks are Proterozoic in their age of metamorphism, belonging to the Grenville tectonic and metamorphic province of the Canadian Shield, which gives ages of 1200–800 Ma. There is a variety of metamorphic and igneous rock types – metamorphic including gneisses, amphibolites and rare marbles, and igneous including gabbros, various felsic rocks and large bodies of anorthosite. The geological map indicates that the metamorphic rocks have undergone several phases of deformation, and that the anorthosite has been intruded (or emplaced) cross-cutting the lithological units of the metamorphic rocks.

The metamorphic rocks are unusually coarse grained (Fig. 8.15), and display a distinctive type of metamorphism, in which the places of hydrous minerals such as micas and amphiboles, are taken by anhydrous minerals such as pyroxene, garnet and aluminium silicates. Since this replacement is likely to have occurred as a consequence of dehydration reactions occurring during prograde metamorphism, we might expect the metamorphism of the Adirondack gneisses to have occurred at unusually high temperatures, and recent research in geothermometry confirms that this is indeed the case in the Adirondacks. Anhydrous regional metamorphic rocks of this type are found in many Archaean and Proterozoic terrains. They indicate a distinctive type of regional metamorphism, which is called 'granulite facies metamorphism'. Before the metamorphic rocks of the Adirondacks, and their conditions of metamorphism are discussed, however, we must say a few words about the name of the **granulite facies** itself.

The granulite facies was one of the original broad categories of metamorphism distinguished by Eskola (1920). He named it after distinctive metamorphic rocks which occur in Saxony, in eastern Germany. These rocks are of felsic igneous composition, rich in alkali feldspar and quartz, and have distinctive metamorphic fabrics which are due to intense deformation of the rocks while they were hot, close to peak metamorphic conditions. The mineral assemblages in many of these rocks show that peak metamorphic conditions reached what is now recognized as the granulite facies of regional metamorphism. However, there has been disagreement almost from the start about how the term **granulite** should be applied to other areas.

Some petrologists have extended the name 'granulite' to metamorphic rocks rich in quartz and alkali feldspar which were not metamorphosed under the distinctive granulite facies conditions. Fortunately, this usage has become obsolete. An alternative which is frequently followed is to extend the name 'granulite' to all kinds of rocks metamorphosed in the granulite facies, giving such names as 'basic granulite'. There is a further complication because of the use of the name charnockite for felsic igneous rocks, metamorphosed in the granulite facies.

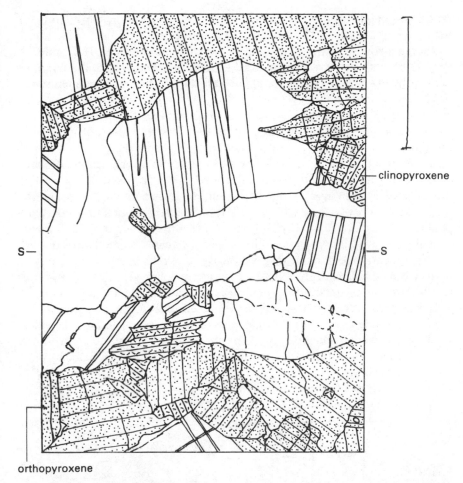

clinopyroxene

s—

—s

orthopyroxene

Figure 8.15 Coarse-grained plagioclase + augite + hypersthene gneiss from Whiteface Mountain, New York State, USA. s–s foliation direction in gneiss. Scale bar 1 cm.

Uniformity of nomenclature among rocks of the granulite facies is thus some way off, although the International Commission on metamorphic rock names may help. Meanwhile, I offer the following guidelines to readers to avoid adding to an already complicated situation. Make a clear distinction between 'granulite facies', which refers to conditions of metamorphism, and 'granulites' and 'charnockites' which are rocks with particular mineral compositions and textures. At least when naming new occurrences of metamorphic rocks, confine the rock names granulite and charnockite to areas where they are already in widespread use, (e.g. Saxony and India respectively). The rocks of the Adirondack Mountains, (to return to our example), are pyroxene gneisses, metamorphosed under granulite facies conditions. The term 'pyroxene gneiss' is descriptive and easy to understand.

Pyroxene gneisses could be formed under other metamorphic conditions than the granulite facies, but this would be rather exceptional.

The gneissose banding is unusually widely spaced in the Adirondacks (Fig. 8.16), with bands several metres in thickness. The dips of the banding are usually low. Geological mapping reveals the presence of several episodes of large-scale folding.

When they are analysed by electron microprobe, the minerals of the Adirondack gneisses turn out to be unzoned. Determination of the distribution of elements between co-existing minerals allows the temperatures of metamorphism to be ascertained. This has been done using a variety of pairs of minerals, and the temperatures calculated can be checked by looking at the mineral assemblages, some of which can co-exist only over certain restricted ranges of temperature. Perhaps because the metamorphic temperatures which are calculated fall within the range of temperatures where the equilibrium stability relationships of minerals can be directly determined by experimental methods in the laboratory, different methods of geothermometry yield impressively consistent results in Adirondack rocks. For this reason, three of the methods will be explained in more detail.

One method of temperature estimation used by Bohlen & Essene (1977) involves a study of plagioclase feldspars and potassium feldspars, which crystallized together when the pyroxene gneisses underwent metamorphism. Figure 8.17 is a phase diagram showing the phase relationships between pure albite ($NaAlSi_3O_8$, Ab) and pure K-feldspar ($KAlSi_3O_8$, Or) at 0.5 GPa

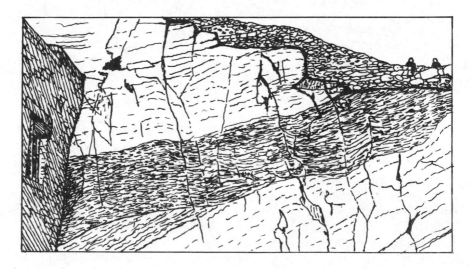

Figure 8.16 Metre-scale banding in gneiss, summit of Whiteface Mountain, Adirondack Mountains. The more pyroxene-rich band of gneiss crossing the foliation in the surrounding gneiss is probably a metamorphosed intrusion.

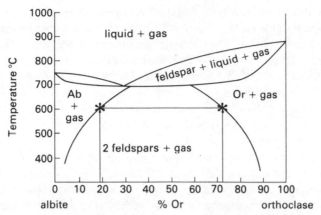

Figure 8.17 Albite–orthoclase phase diagram, at 0.5 GPa pressure. The compositions of Na-rich and K-rich feldspars co-existing in equilibrium during metamorphism (asterisks) give the temperature of metamorphism as 600°C.

pressure. At any temperature T below the melting temperature T_m, plagioclase feldspar which crystallized in equilibrium with K-feldspar will contain an appreciable proportion of potassium, and the K-feldspar an appreciable proportion of sodium. The differences may be expressed as proportions of the end-members Ab and Or. For example, at the temperature T_1 of 600°C, illustrated in Figure 8.17, the plagioclase composition is $Ab_{81}Or_{19}$ and the composition of the K-feldspar is $Ab_{25}Or_{25}$. Because the phase boundaries between the two-feldspar field and the albite and K-feldspar fields slope, determination of the compositions of co-existing feldspars will vary considerably with temperature. The plagioclase–K–feldspar system is said to be a useful geothermometer.

There are complications to applying this method to the Adirondack gneisses. One is that the plagioclase feldspars are usually not pure albite, but this can be allowed for because the equilibrium stability relationships of feldspars have been well calibrated experimentally. A more serious problem arises over the ordering of Si and Al ions onto different sites within the crystal structures of feldspars, but this has been resolved by thermodynamic calculation methods in the temperature range we are considering here.

There are also practical problems. The Adirondack gneisses cooled slowly from their peak temperature, and during this time the large feldspar crystals continued to adjust their compositions to equilibrium values at lower temperatures. For example, at a temperature of 500°C in Figure 8.17, the K-feldspar, which had the composition $Ab_{25}Or_{75}$ will have altered to K-feldspar of composition $Ab_{18}Or_{82}$ and albite of composition $Ab_{90}Or_{10}$. At these lower temperatures, fortunately, the albite produced by this readjustment remains inside the former K-feldspar crystals, giving rise to the type of mineral intergrowth known as **perthite**. The relative proportions of

K-feldspar and albite can be calculated from the diagram using the lever rule, they are K-feldspar 92%, albite 8%. What Bohlen and Essene did was the opposite calculation. They determined the relative proportions of K-feldspar and albite in the perthite, and of albite and K-feldspar in the anti-perthite crystals of the same rock. They also determined the compositions of albite and K-feldspar, using the electron microprobe. They were then able to *re-integrate* the original compositions of the perthite and anti-perthite crystals, and apply these calculated compositions to work out the temperature of metamorphism.

Notice that this method yields more than one temperature. The compositions of K-feldspar and albite in perthite yield the lower temperature T_2, and the re-integrated compositions of perthite and anti-perthite the higher temperature T_1. These temperatures were achieved at different times t_1 and t_2 on the P–T–t path (Fig. 8.18). At time t_1, what happened was that sodium and potassium ions ceased to be exchanged between the large K-feldspar and albite crystals. At t_2, sodium and potassium ceased to be able to diffuse

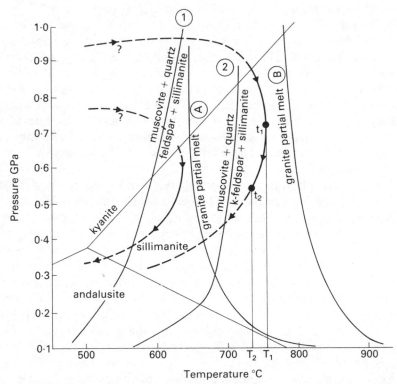

Figure 8.18 Part of two P–T–t paths for Adirondack gneisses. The earlier parts of the P–T history cannot be deciphered, because the high temperature granulite facies metamorphism has destroyed any parts of the minerals which might have survived to preserve earlier stages. For significance of times t_1 and t_2, and temperatures T_1 and T_2, see text.

through the crystal lattices of the feldspars. Both are blocking temperatures (Dodson 1973), at which some exchange process ceased to be possible. Temperatures such as T_1, at which exchange of chemical components ceases between the large crystals constituting the principal metamorphic fabric of the rock, are usually the temperatures which are described as 'temperatures of metamorphism' in metamorphic geothermometry. Because such exchange processes are thermally activated, it is quite likely that temperatures such as T_1 are the maximum metamorphic temperatures, but it is by no means certain. The lower temperature T_2 was achieved later, because the perthite and anti-perthite separated from crystals which were previously homogeneous, at temperature T_1. The rocks of the Adirondack Mountains preserve the record of the retrograde part of the P–T–t paths in a wide variety of textures, but the prograde part has been obliterated by the widespread attainment of equilibrium about time t_1. This state of affairs is often found in rocks from the deeper parts of the continental crust.

Heat flow and thermal gradients in continental crust

We have seen that in long stabilized areas of continental lithosphere, the heat flow tends to settle down to a value of about 35 mW, and the geothermal gradient to the steady-state geotherm. In the early history of the Earth, the rate of production of heat in its interior must have been greater, because radioactive isotopes were more abundant and so their decay produced more heat. If heat production in the Earth was greater, geothermal gradients should have been higher, in the stable parts of plates where heat escaped through the whole thickness of the lithosphere by conduction.

The observation that granulite facies rocks are far more common in Precambrian shields than in Phanerozoic metamorphic belts lends some support to this idea. Granulite facies metamorphism requires that temperatures become high at relatively shallow depths, but it also needs a process to prevent melting of crustal rocks when the minimum melting temperature of granitoid magmas is exceeded. There are two possible mechanisms for suppressing melting, both depleting the rocks in H_2O, and thus raising the melting temperature. One is that the H_2O had been driven out during an earlier metamorphic cycle (Thompson 1983), the other that the fluid present during metamorphism was CO_2 rather than H_2O (Hansen *et al.* 1984). Both mechanisms could have occurred in the earliest stabilized parts of the lithosphere (Crawford & Hollister 1986). Granulite facies rocks are not abundant in the oldest Archaean metamorphic areas (Windley 1984), where the rocks are migmatites and schists showing Barrow-type regional metamorphism. The middle to late Archaean saw the most widespread development of granulites, accompanied by large intrusions of basic and ultrabasic magmas.

A possible explanation is that the remnants of metamorphic rocks which survive from early Archaean times, crystallized during a period when tectonic plates were smaller, and rates of plate movement greater, than they are today, so that geothermal gradients were not able to relax to a steady-state geothermal curve anywhere at all. In mid-Archaean times, for the first time, areas of lithosphere remained stable over periods of 10–100 Ma, and high geothermal gradients became stabilized in them. The upper mantle was extensively melted, forming large volumes of basic and ultrabasic magma, and granulite facies metamorphism took place over large areas of the crust. In Proterozoic and Phanerozoic times, the global rate of heat production fell, and the combination of high geothermal gradients and long-stabilized lithosphere became unusual.

9 Metamorphic rocks of the mantle

In certain unusual geological environments rocks derived from the Earth's mantle are found at the surface. They have come there in one of two ways, either by tectonic movements or by volcanic eruption. Many mantle rocks coming to the surface in either way can be recognized as metamorphic by their textures.

The density of the mantle and the velocity of transmission of seismic waves through it suggest that it has a predominantly ultrabasic igneous composition. However, there are rocks of basic igneous composition as well, and examples of both compositions will be described here.

The deepest parts of ophiolite complexes, which were discussed in Chapter 5, are of ultrabasic composition, and are regarded as the uppermost part of the mantle, underlying oceanic crust produced by extension of an oceanic ridge. This uppermost part of the mantle has usually had its composition modified by partial melting of the parent mantle rock, followed by removal of the basaltic melt upwards to form the oceanic crust (Gass & Smewing 1981, Brown & Mussett 1981). The rocks are peridotites of the **harzburgite** division (olivine + orthopyroxene), and they usually show a coarse-grained foliation marked by a preferred elongation of prismatic orthopyroxene crystals and often also a preferred orientation of the olivine crystals. The chromium-rich spinel which is an important minor mineral in the rock may also pick out the foliation, either by being concentrated into layers, or by the originally octahedral cubic crystals being broken and stretched out into streaks. The foliation has a gentle dip, being folded into the same anticline as the crustal part of Troodos Ophiolite of Cyprus, and running approximately parallel to the individual layers of oceanic crust.

Geochemical studies on harzburgites and basic rocks from the Troodos Complex show that the harzburgites were once the solid residue left behind in the Upper Mantle after an appreciable proportion of basalt of MORB composition had been removed following partial melting. The basalt escaped up fractures beneath the oceanic ridge to form the oceanic crust, leaving behind the mantle peridotite, still solid. The textures of the peridotite were produced by deformation in the solid state, which occurred when the mantle material welled up beneath the oceanic ridge, then rotated to flow horizontally away from the ridge beneath the ocean floor (Fig. 5.8). Many ophiolite complexes retain fragments of mantle from beneath the oceanic crust, as at Troodos. This mantle material is clearly derived from the shallowest level within the Earth's mantle.

To find the nature of deeper mantle rocks, we must look elsewhere. Mantle material from beneath both oceanic and continental crust is brought to the surface as **xenoliths** in lavas of volcanic areas. They occur most frequently in basaltic igneous rocks, but also in the unusual volcanic rocks that are the source of diamonds (Nixon 1987). These are called **kimberlites** after the famous Kimberley diamond-mining district of South Africa. The blocks in both basalts and kimberlites include rocks of ultrabasic and basic composition. There are examples with both igneous and metamorphic textures.

Kimberlites are fragmental rocks with different types of fragments in a fine-grained matrix of low-temperature minerals such as serpentine and calcite. The blocks include the country rocks which surround the kimberlite at the surface but also exotic blocks which have come from deeper parts of the crust and from the upper mantle. Kimberlites are only found in Precambrian shield areas of continental crust, or in areas where such crust is covered beneath Phanerozoic sediments. They outcrop over small areas, often of irregular shape, but mining shows that the kimberlite bodies are usually vertical pipe-like structures cutting through the Precambrian rocks. Associated small dykes and sills are occasionally found. The average diameter of the pipes in the typical Kimberley district, is 300 m increasing gradually upwards so that the pipe has the shape of an inverted cone. Their structure therefore suggests that they are volcanic vents, and this is supported by the discovery of erupted volcanic rocks resembling kimberlites in eroded volcanoes in Tanzania and Namibia (Reid *et al.* 1975, Ferguson *et al.* 1975).

Kimberlites are therefore probably volcanic rocks formed by eruption from deep in the mantle, the material erupted at the surface being a mixture of gas and solid fragments torn from the walls of the vent. The gas must have originated in the mantle and its explosive escape upwards brought rocks from very deep in the Earth to the surface, so that the xenoliths in the kimberlite include rocks from the mantle and from the crust. Ultrabasic blocks from the mantle are commoner, but basic rocks also occur. Both compositional types may display foliation and lineation, and other metamorphic textures, particularly those produced by dynamic metamorphism (Boullier & Nicholas 1975). They are appreciably coarser grained than crustal metamorphic rocks, with grain sizes up to ten times greater.

Figure 9.1 shows the hand specimen of a rock of basic igneous composition from the Roberts Victor kimberlite pipe, South Africa (Lappin & Dawson 1975, Harte & Gurney 1975). This pipe is unusual because the xenoliths in it are mainly basic. The major minerals can easily be recognized as garnets up to 10 mm in diameter, and pyroxene crystals of comparable size (the garnets are deep red and the pyroxene is pale green). Although the rock is so coarse grained, it has an even granoblastic texture. Under the petrological microscope the simple primary mineral assemblage is confirmed as garnet + clinopyroxene. The rock has a basaltic composition, and is therefore an eclogite.

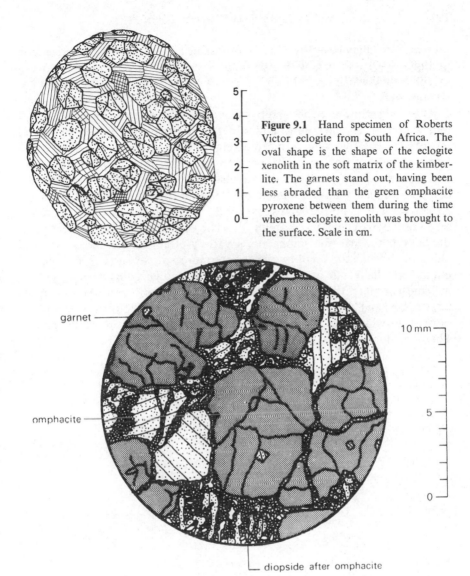

Figure 9.1 Hand specimen of Roberts Victor eclogite from South Africa. The oval shape is the shape of the eclogite xenolith in the soft matrix of the kimberlite. The garnets stand out, having been less abraded than the green omphacite pyroxene between them during the time when the eclogite xenolith was brought to the surface. Scale in cm.

garnet

omphacite

diopside after omphacite

Figure 9.2 Thin section of eclogite block from the Roberts Victor kimberlite pipe, South Africa. Scale bar 10 mm.

Eclogites have an average density of about 3500 kg m^{-3} whereas gabbros of the same composition have a density of about 3000 kg m^{-3}. The higher density suggests that eclogite is a rock which crystallized under high pressure, and this caused the ions in the mineral crystals to become more closely packed together and thus denser. Mineral assemblages of calcic plagioclase + clinopyroxene (like basalt) have been changed into garnet + Na bearing clinopyroxene (like eclogite) in laboratory experiments. Eclogite is also formed by crystallization of basaltic magma under high pressure, so it may be either a metamorphic or an igneous rock. The eclogite in Figures 9.1 and 9.2, with its granular texture, may be either a metamorphic rock which crystallized under stress-free conditions, or a granular igneous rock.

At the edges of the garnet and pyroxene crystals there are reaction rims of much smaller crystals, and undoubtedly these have formed by metamorphism. The pyroxene is more extensively altered than the garnet, the large crystals of parent pyroxene being replaced along boundaries and cracks by smaller crystals, which are also of pyroxene, but with a different composition. They have a similar birefringence, but a lower R.I. than the parent crystals. The parent pyroxenes have $2V_\gamma$ of about 80°, the small secondary pyroxenes $2V_\gamma$ of about 30°. The optical properties indicate that the large pyroxenes are omphacite, a variety of augite rich in sodium and aluminium, while the small pyroxenes are more ordinary augite. Omphacite is a mineral diagnostic of eclogites and may be recognized by its green colour in hand specimen (possibly visible also under the microscope) and by its large optic axial angle. Garnet + pyroxene rocks also occur (for example in skarns), in which the pyroxene is not omphacite, but diopside or acmite. These are not eclogites and did not form at unusually high pressures.

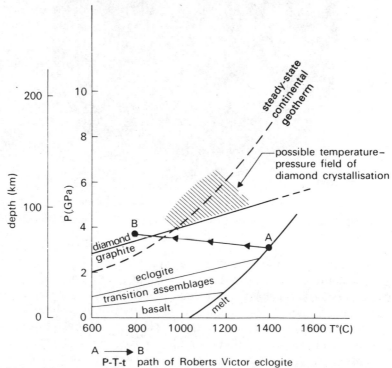

Figure 9.3 P–T–t path of Roberts Victor eclogite, from Lappin and Dawson (1975).

The association of eclogite with diamonds in kimberlites shows that they formed at high pressures and therefore at considerable depths. Diamonds have been found in one or two specimens of eclogite, and inclusions of eclogite and its constituent minerals have been found in diamonds, so their coexistence under the conditions of metamorphism in the mantle has been adequately demonstrated. Figure 9.3 shows the univariant curve for the reaction

$$graphite = diamond$$

which has been well established experimentally. The depths equivalent to the pressures are also given. The diagram shows that if diamond + omphacite + - garnet is correctly identified as an equilibrium metamorphic assemblage, the metamorphism must have occurred deep in the mantle.

By studying the **retrograde** metamorphic reactions, such as those at the boundaries of the garnet and pyroxene crystals in Figure 9.2, it has proved possible to determine the retrograde part of P–T–t paths which the eclogite xenoliths followed in the mantle (the high pressures revealed by the compositions of co-existing pyroxene and garnet show that the paths were followed in the mantle, not during the pressure release which accompanied the eruption of the kimberlite). Individual blocks of eclogite show different P–T–t paths, which presumably reflect their different depths in the mantle when they were caught up into the kimberlite.

It is the changes in the amount of aluminium in the clinopyroxene which reflect changes in metamorphic temperatures and pressures particularly strongly. They are shown in ACF diagrams in Figure 9.4, which also shows that garnets may be much richer in Mg (i.e. the pyrope end-member composition) than those of crustal regional metamorphic sequences.

The gap in composition between co-existing clinopyroxene and orthopyroxene becomes wider from the early compositions shown in triangle (a) and the later compositions shown in triangle (c). This reflects an increasing Ca content in clinopyroxene, and an increasing Mg, Fe content in orthopyroxene, i.e. each pyroxene is coming nearer to its ideal or stoichiometric composition. There is a solvus gap in pyroxene compositions, and the reduction in the mutual solution of the pyroxenes mainly reflects a *fall* in temperature from (a) to (c). The change in the Al content of the pyroxenes is more complex because it reflects a change in mineral assemblages from kyanite + omphacite in (a), to kyanite + garnet + omphacite in (b), to garnet + diopside in (c). This sequence reflects an *increase* in pressure, as the temperature decreased.

If a speculative prograde path is added to the retrograde P–T–t path of Figure 9.3, it can be seen that the resultant P–T–t loop is *anticlockwise*, in contrast to the clockwise P–T–t loops which we saw in crustal regional metamorphic rocks from collision belts. The pressures recorded in the

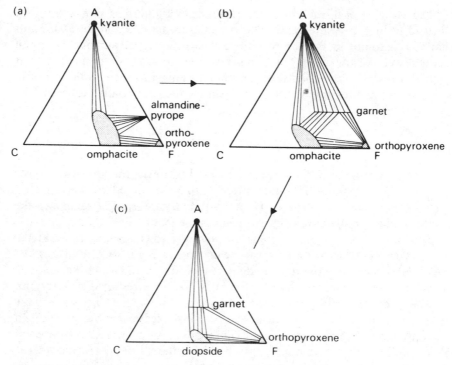

Figure 9.4 ACF diagrams showing the sequence of stable minerals assemblages in the Roberts Victor eclogites.

mineral assemblages indicate that metamorphism of the eclogite blocks occurred at depths of 100–200 km in the mantle, in the **asthenosphere** where mantle rocks are particularly plastic, and it may be that the form of the P–T–t loop is related to the movements of mantle material beneath continental lithospheric plates.

Eclogites do not occur only in the mantle. They occur also in high-grade gneisses, usually as pods only a few metres across, but occasionally as larger bodies also. They are also characteristic of the higher-grade parts of areas of metamorphic rocks which have undergone subduction, such as those in the Pennide Alps. Such rocks also indicate very high pressures, though not so high as those recorded in mantle derived eclogites. Although it is eclogites which have been described in this section, peridotites are commoner as mantle-derived xenoliths, and it is peridotite composition which will be considered when we speculate about the nature of rocks at greater depths in the mantle.

At a depth of 670 km beneath both continents and oceans, there is a relatively sharp change in the physical properties of the rocks of the Earth's mantle, revealed by seismological studies (Brown & Mussett 1981, Bott 1982). Below this depth, the mantle rocks are more uniform in their properties until the base of the mantle is reached at a depth of 2886 km. Deeper still lies the Earth's fluid metallic core. The 670 km discontinuity in the mantle is therefore regarded as the upper surface of the Lower Mantle. As far as we know at present, rocks from the Lower Mantle have not reached the surface of the Earth in an unmodified state, so that we can study them as we do rocks from the Upper Mantle and Crust, but high pressure experiments and calculations of mineral stability fields have recently permitted research workers to calculate which minerals probably form the rocks of the Lower Mantle, and thus enter the realm of well-founded speculation, rather than wild guesswork. Because the Mantle makes up the majority of the Earth's total mass, these unfamiliar minerals, if they have been correctly identified, are the commonest on Earth.

The 670 km discontinuity is apparently the result of a **phase change** in the minerals of the mantle, caused by increasing pressure, rather than a change in the chemical composition of the mantle (Brown & Mussett 1981, Bott 1982). It is similar to the phase change between gabbro and eclogite shown in Figure 9.3, but occurs at a much higher pressure, and in material of ultrabasic composition, rather than basic. The 670 km discontinuity in the mantle is the deepest of several which have been discovered by seismologists at depths between 200 and 670 km, in the part of the mantle known as the Transition Zone. Several of these discontinuities are attributed to phase changes in the mantle peridotites, and the proposed sequence is summarized in Figure 9.5 (Jeanloz & Thompson 1983).

The most significant changes are those which occur in material which has the composition of olivine $(Mg, Fe^{+2})_2SiO_4$ in the crust and upper mantle. The change occurring at the lowest pressure is that silicon (Si^{+4}) changes from being surrounded by four oxygen ions (O^{+2}) as in olivine, to being surrounded by six oxygen ions. This produces a mineral with a structure similar to spinel $MgAl_2O_4$, which because the oxygen ions are more closely packed together, is denser than olivine. At shallow levels, the crystal struc-

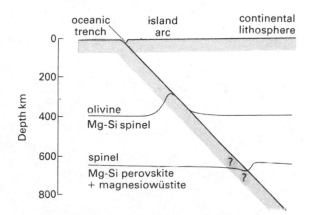

Figure 9.5 Discontinuities and associated phase changes in the upper mantle near a subduction zone (same scale as Figure 6.1). From Bott (1982).

ture is distorted compared with true spinel, so that it has orthorhombic not cubic symmetry. This mineral is known as the beta phase. With increasing depth (or pressure) it changes to a mineral with a cubic crystal structure, which is known as the gamma phase.

However, the 670 km discontinuity is probably the result of a further phase change. The stable magnesium silicate has the structure of the mineral perovskite ($CaTiO_3$), with some magnesium ions surrounded by 12 oxygen ions (as the Ca ions are in perovskite). For the olivine-rich mantle rock compositions, this leaves an excess of MgO and FeO, which crystallize as the dense oxide mineral magnesiowüstite. The equation for this reaction is

$$(Mg, Fe^{+2})SiO_4 = (Mg, Fe^{+2})SiO_3 + (Fe^{+2}, Mg)O$$

olivine   Mg-perovskite   magnesiowüstite

If these two minerals are correctly identified as those which form most of the lower mantle, they are the most abundant minerals on Earth.

Subducted lithospheric plates descend to as much as 700 km depth, and therefore the phase transformations just described, olivine to beta phase, beta phase to gamma phase, and gamma phase to Mg–perovskite + magnesiowustite may occur in the mantle part of the plate as it descends. In this setting, the reactions would occur in the solid state, and so these phase changes are metamorphic reactions. The subduction of lithosphere is a process of great significance for the fundamental mechanism of plate tectonics (which is still far from understood), and thus metamorphism assumes a major role among Earth processes. It would be fascinating to obtain rocks from the deeper parts of subducted lithospheric plates for petrological study.

10 Extraterrestrial metamorphism

Impact metamorphism

The types of metamorphism so far described in this book have been dominated by deformation processes due to tectonic movements within the Earth due to the relative motions of lithospheric plates. There is a different type of metamorphism, which is due to intense, rapid deformation following the impact of extraterrestrial bodies onto the Earth's surface. Particles of extraterrestrial matter (**meteoroids**) arrive regularly in the vicinity of the Earth. They enter the atmosphere with velocities in the range 5–20 km s^{-1}, due to the effects of relative orbital velocities, plus acceleration in the Earth's gravitational field. Interaction with the atmosphere sorts the particles of extraterrestrial matter according to their size – small particles are burned up by friction before they strike the ground (producing the incandescent trails in the night sky called **meteors**), particles of intermediate size are slowed down so that fragments fall to the surface relatively undamaged (as **meteorites**), while the largest particles are only slightly slowed down, and strike the surface at high speed. It is the last category, fortunately the rarest, which produces shock metamorphism.

The high-speed impact of a massive meteoroid, weighing 10^8 kg (100 tonnes) or more, produces a violent explosion, in which the matter in the meteoroid itself is fragmented and vaporized. The shock waves travelling outwards through the surrounding rocks from the impact site cause metamorphism as well as a variety of distinctive structures. In contrast to contact metamorphism, dynamic metamorphism associated with tectonic movements, and regional metamorphism, shock metamorphism takes place extremely quickly, during a few microseconds to seconds. In this brief interval, temperatures of several thousand °C, and pressures in excess of 10 GPa (100 kbar) may be reached.

Probably the most famous meteoroid impact site in the world is Meteor Crater, Arizona, USA. The meteoroid fell approximately 15 000 years ago onto flat-lying Permian to Triassic rocks of the Colorado Plateau. In the desert climate, the shape of the crater has only been modified a little by subsequent erosion and deposition of sediments in the bottom of the crater (which was occupied by a lake at one time). Figure 10.1 shows a cross section through the crater and its surrounding ejecta blanket.

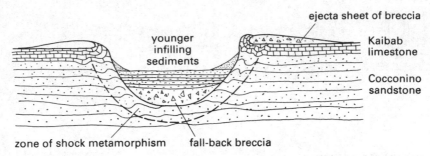

Figure 10.1 Cross section through Meteor Crater, Arizona, USA. Note overturning of Kaibab limestone strata on the rim of the crater.

The intense shock waves fragmented the rocks which were ejected from the crater, so that the ejecta consists of **breccia** of rock fragments. It is these fragments which show the effects of shock metamorphism, which are most characteristically seen in the metamorphosed sandstone. The shock waves transformed some of the quartz in the sandstone into the high-pressure **polymorphs** of SiO_2, coesite and stishovite, both of which were first identified as natural minerals from Meteor Crater. Even more intense shock metamorphism has transformed the quartz grains into glass of SiO_2 composition. This glass has a higher density and R.I. than the glass obtained by melting and quenching quartz, and is therefore given a special name, lechatelierite.

Shock metamorphism at the Nördlinger Ries Crater, Germany

The Ries Crater is a remarkable, almost circular hollow 21–24 km in diameter. It lies near the western border of the state of Bavaria in southern Germany, between the Franconian and Schwabian Hills (Fig. 10.2). The ancient city of Nördlingen stands on the Neogene sediments which fill the hollow.

The origin of the structure has been the topic of considerable discussion which is now largely resolved. It is generally agreed to be a somewhat eroded meteoroid impact crater. Figure 10.2 shows an east–west cross section through the structure. The local geological succession is relatively simple, with flat-lying Mesozoic sediments unconformably overlying a complex of metamorphic and igneous rocks which formed during the late Palaeozoic Variscan orogeny. The Variscan basement complex consists mainly of biotite granite containing dykes of amphibolite. The Mesozoic sediments include Keuper marls, Liassic clays and a thick limestone of late Jurassic age. The succession within the crater is very different (Fig. 10.2). Beneath Neogene sediments is a remarkable deposit which consists of a jumble of blocks of biotite granite and the sediments found above it outside the crater. This blocky deposit rests directly on the crystalline basement complex, with a strong discordant contact at the surface between the two.

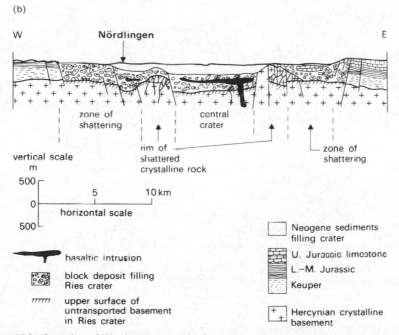

Figure 10.2 Location of Ries impact crater in Germany, and cross section through the crater.

The meteoritic impact origin of the Ries Crater has been demonstrated from petrographic and mineralogical studies on the crystalline basement rocks affected by shock metamorphism. In the hills surrounding the crater, a sheet of material ejected during its formation is found in the Neogene sedimentary succession. The position of the sheet in the stratigraphical succession fixes the age of formation of the crater between the beginning and middle of the late Miocene (about 1.5 Ma). The ejecta sheet consists of a tuff-like rock called **suevite**, containing larger ejected blocks of all the country rocks. The granite blocks show a complete gradation from relatively unaltered granite, to granite completely transformed by metamorphism into glass. They have been arranged into a sequence, showing successive stages of progressive shock metamorphism, by Stoffler (1966).

Figure 10.3 shows a thin section of a fine-grained specimen of suevite. The matrix material is brown glass, containing numerous vesicle-like holes, lined with darker brown glass. The fragments in the suevite are of several kinds, including marble, clear glass, K-feldspar, plagioclase feldspar, biotite and quartz. The suevite should strictly be regarded as an extremely unusual type of *sedimentary* rock, composed of shock metamorphic fragments, ejected from the crater, in a glass matrix.

The feldspar porphyroclasts are especially interesting. Originally, the clast in the middle was probably composed entirely of K-feldspar. Its interior is of colourless isotropic glass, and its outer part of radiating crystals of K-feldspar, which appear to have crystallized from the glass. The R.I. of the

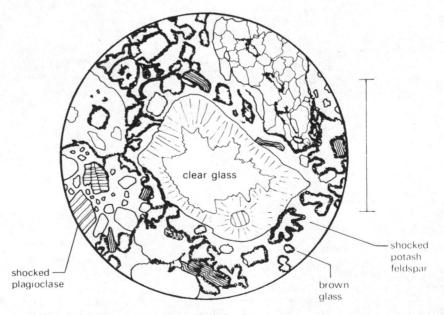

Figure 10.3 Suevite from the Ries Crater. Scale bar 1 mm.

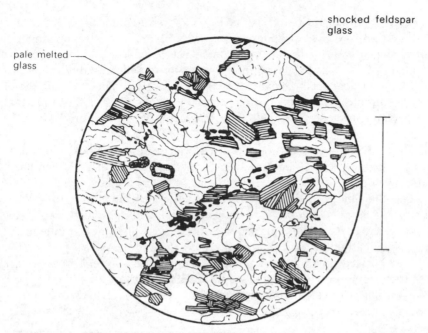

Figure 10.4 Shock metamorphosed granite from an ejected block, Ries Crater. Scale bar 1 mm.

glass is *higher* than that of the K-feldspar, which indicates that it is equivalent to the lachatelierite glass derived from quartz by shock metamorphism. The constituent ions of the K-feldspar structure were shaken from their regular positions by the intensity of the shock waves produced in the impact, and fell back into random positions. This process can be simulated in the laboratory, and only shock waves formed by very powerful, short-lived explosions are intense enough to do it. Underground nuclear explosions are intense enough, and so are impacts by projectiles travelling above the velocity of sound in rock, but volcanic explosions are probably not. Another interesting clast is of plagioclase feldspar, whose twin lamellae have been displaced into three separate zones, separated by planar fracture surfaces. Biotite clasts have had their cleavage planes bent into kink bands. Although a few of these microstructures may also be produced by tectonic deformation, the total range of transformations can only be explained by shock metamorphism.

Stoffler (1966) has distinguished four stages in the shock metamorphism, as follows:

low grade   I   Development of planar cleavage cracks in quartz.

             II   Partial-to-total transformation of feldspars to glass.

             III   Reduction of pleochroism and birefringence in biotite.

high grade IV   Shock metamorphism of all minerals in the granite.

In Figure 10.4 the feldspar clasts have been shock metamorphosed to stage II. The biotite clasts, although deformed, retain their pleochroism and birefringence, showing that metamorphism was below stage III. But the clast at upper left is of clear glass containing vesicles, and probably represents quartz or feldspar metamorphosed to stage III or IV. Thus one sample of suevite contains fragments of different metamorphic grades, indicating that the metamorphism occurred before the fragments were incorporated into the breccia.

Figure 10.5 shows a thin section from an ejected granite block. The quartz and feldspars have been entirely transformed into clear glass, which has a pattern of cracks similar to perlitic cracks in volcanic glass. Veins of glass traverse the slide, which were formed by melting of the granite to a liquid during the shock metamorphism. In plane polarized light, as shown in the drawing, the parts of the thin section without veins look like granite under low magnification, because they retain the texture of the parent granite, but between crossed polars all the crystals except the biotite and opaques are seen to be composed of isotropic glass. The biotites show kink banding, but have normal pleochroism and birefringence, and the rock is therefore from the high-grade part of stage II.

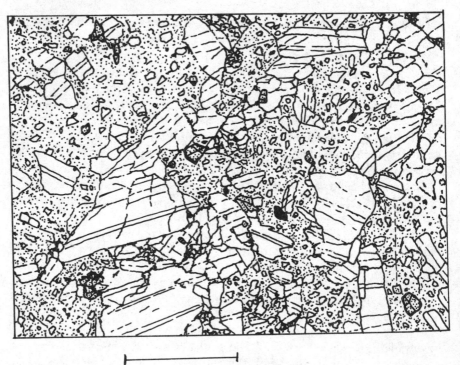

Figure 10.5 Brecciated anorthosite, Descartes Highlands region, the Moon. Apollo sample number 60025,159. Scale bar 1 mm.

The effects of shock metamorphism are so distinctive in thin section that there can be no doubt that the fragments thrown out of the Ries Crater underwent metamorphism by shock waves produced by an extremely powerful and very brief explosion. Meteorite impact is the only natural phenomenon which meets the case.

Metamorphism in Moon rocks

Because the Moon has no atmosphere, meteorites strike its surface at high velocities (8–12 km s^{-1}). Shock metamorphism is therefore seen in virtually all Moon rocks while the meteorites themselves have not been found because the smaller ones were totally vaporized upon impact. Matter from them has been detected by geochemical and mineralogical study of the loose layer of 'soil', forming the uppermost 3 m or so of the Moon's crust. This layer is more correctly called **regolith**. Rocks of granitoid composition are extremely rare on the Moon and so direct comparison of Moon rocks with the shock metamorphosed granite of the Ries Crater is not possible.

The highland regions of the Moon (which are the lighter areas seen from the Earth) are composed entirely of breccia. These breccias resemble the suevite and block deposits of the Ries Crater in their textures, but have different compositions. The clasts consist of basic igneous rocks, both basalts and coarser-grained gabbros, and of anorthosite. They include the oldest rocks found on the Moon.

Figure 10.5 shows a thin section from an anorthosite clast from the highland region of the Moon. The entire section comes from part of a large anorthite crystal, which has been fractured by shock metamorphism. The less disrupted part of the crystal shows twin lamellae of plagioclase feldspar displaced along fractures produced by shock, while in the more disrupted part the crystal has been broken into extremely poorly sorted plagioclase clasts. Examination under higher magnification shows that the matrix between the clasts contains smaller plagioclase fragments, the whole cemented by a glass of plagioclase composition.

Figure 10.6 shows a highland breccia with clasts of different types. Like the Ries Crater suevite, this breccia is strictly a sedimentary rock fragmented by meteorite impact and transported as the ejecta sheet. The clasts show the effects of intense shock metamorphism. This particular breccia is derived from the thick sheet of ejecta (the Fra Mauro Formation). This ejecta was thrown out of the immense Imbrium circular basin following the impact of a large meteoroid on the Moon's surface 3900 Ma ago. The sample was collected by the Apollo 14 expedition to the Moon in 1972, which was specially planned to discover the nature of the Fra Mauro Formation.

When the samples brought back from the Apollo 14 mission were studied, it was found that the matrix material of some breccias has undergone

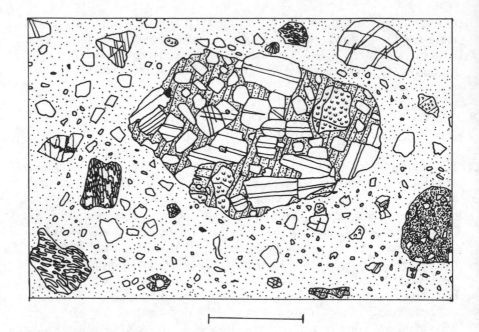

Figure 10.6 Polymict breccia with metamorphic clasts, Taurus–Littrow region, the Moon. Apollo sample number 73235,63. Scale bar 1 mm.

Figure 10.7 Clast in metamorphosed breccia, showing granoblastic texture due to earlier metamorphism. Taurus–Littrow region, the Moon. Apollo sample number 79215,67. Scale bar 1 mm.

metamorphism. Lunar highland breccias may be classified into three groups by the nature of their matrix – (1) breccias with a matrix which preserves original impact-melted glass, (2) breccias with a metamorphic matrix and (3) breccias whose matrix has crystallized slowly from a melt. Figure 10.6 shows one of the breccias with a metamorphic matrix. Under high magnification, the matrix can be seen to be made up of crystals of plagioclase and pyroxene, showing a granoblastic texture. The metamorphism of the matrix took place during the cooling of the hot Fra Mauro Formation; shallower levels cooled more quickly, yielding the breccia retaining glass fragments in its matrix, deeper levels remained hot for longer, so that the matrix was melted and then crystallized as large grains, often including non-melted clasts of plagioclase and spinel. By studying different specimens of Fra Mauro breccias, it was possible to identify a progressive metamorphic sequence in the matrix material (Warner 1972). The outer layers of the ejecta sheet had very low **thermal conductivity**, permitting the interior to remain hot for long enough for the matrix to become metamorphosed, or at deeper levels to melt and subsequently crystallize. The metamorphism took place at relatively high temperatures (> 800°C), and Warner described it as 'lunar thermal metamorphism', because although the temperature was high, the depth of burial was comparatively low. Of course it is not contact metamorphism, because the high temperatures were not brought about by heating due to the intrusion of igneous magma. Figure 10.7 shows a breccia with a metamorphic clast in a granoblastic matrix.

This lunar metamorphism is of interest for metamorphic processes on Earth because there is no H_2O in any Moon rocks, and there probably was none at the time of metamorphism either. Moon rocks therefore show that metamorphism can occur in the absence of water or other metamorphic fluids.

Appendix

Table of molecular weights to use in calculations of rock and mineral compositions on composition–assemblage diagrams.

Al_2O_3	101.82	MnO_2	86.93
B	10.81	Mn_3O_4	228.79
B_2O_3	69.60	Na_2O	61.97
BaO	153.33	NiO	74.70
BeO	25.01	Nb_2O_5	265.78
C	12.01	P_2O_5	141.92
CO_2	44.00	PbO	223.18
CaO	56.07	Rb_2O	186.93
CeO_2	172.11	S	32.06
Ce_2O_3	328.22	SO_3	80.05
Cl	35.45	Sc_2O_3	137.89
CoO	74.93	SiO_2	60.07
Cr_2O_3	151.97	SnO	134.68
CuO	79.53	SrO	103.61
F	19.00	Ta_2O_5	441.87
FeO	71.84	ThO_2	264.03
Fe_2O_3	159.68	TiO_2	79.89
H_2O	18.01	UO_2	270.02
HfO	210.48	U_3O_8	842.04
K_2O	94.20	V_2O_6	181.85
La_2O_3	325.80	Y_2O_3	225.79
Li_2O	29.87	ZnO	81.36
MgO	40.31	ZrO_2	123.21
MnO	70.93		

Glossary

ACF diagram Triangular composition-assemblage diagram especially useful for basic igneous rocks. The three components plotted are:
A – [Al_2O_3], C – [CaO], F – [$FeO + MgO$] (with various corrections).

AFM diagram Triangular composition-assemblage diagram especially useful for pelitic rocks. The three components plotted are:
A – [Al_2O_3], F – [FeO], M – [MgO] (with various corrections).

amphibolite A metamorphic rock composed mainly of amphibole and plagioclase feldspar in about equal proportions. Usually a medium to high grade regional metamorphic rock of basic igneous composition.

amphibolite facies Moderate P, moderate T facies of regional metamorphism, covering the higher **metamorphic grades** of Barrow type **progressive regional metamorphism**.

asthenosphere The plastic layer within the Earth, underlying the **lithosphere**.

autochthon The part of a mountain chain which has not been displaced tectonically, the tectonic basement.

Barrow type regional metamorphism A type of **progressive metamorphic sequence** commonly occurring in collisional orogenic belts, typified by the sequence of metamorphic zones in the Grampian Highlands of Scotland, described by G. Barrow.

basement The metamorphic rocks underlying the shelf sedimentary sequences of stable continental areas.

black smokers Chimney-like structures, several metres high, on the ocean floor near oceanic ridges. They occur where springs return hot mineralized water from hydrothermal circulation systems in oceanic crust back to the cool ocean water.

blastomylonite A rock formed by **dynamic metamorphism** with a crystalline matrix between **porphyroclasts**.

blocking temperature The temperature below which **components** cease to be able to diffuse freely into and out of metamorphic minerals.

blueschist Schist of basic igneous composition, containing glaucophanitic amphibole, lawsonite, albite and quartz.

blueschist facies High P, low T facies of regional metamorphism, associated with **subduction zones**.

boudin Lenticular body of rock, formed by fracturing and extension of a once continuous layer.

breccia   A rock composed of angular fragments in a finer-grained matrix.

buffered   See **externally buffered** and **internally buffered**.

cataclasis   The process of breaking of mineral grains or rock fragments into smaller fragments during dynamic metamorphism.

cataclastic metamorphic rocks   See **dynamic metamorphic rocks**, which is the preferable term.

charnockite   Metamorphic rocks with granitoid compositions, containing pyroxene in place of mica or amphibole, and usually dark in colour. Metamorphosed under **granulite facies** conditions.

chiastolite   A textural variety of andalusite with dark inclusions arranged in a cross shape within the prismatic crystals.

cleavage   A tendency to split into parallel sheets, in either minerals or rocks. It is strictly known as *rock cleavage* when it is developed in rocks (e.g. slates).

complex   Lithostratigraphical term for a unit of rocks which can be distinguished and mapped in the field, but which cannot be regarded as defined by sedimentary deposition processes. Particularly useful for naming mappable units of metamorphic rocks which are not lithostratigraphical **formations** or **groups**. e.g. 'The Malvern (metamorphic) Complex'.

component   A limiting chemical composition which may be used to express the composition of **phases** in a chemical **system**, e.g. the composition of a pyroxene phase in a pyroxene hornfels may be expressed as 68% of the component $CaMgSi_2O_6$ (diopside) and 32% of the component $CaFeSi_2O_6$ (hedenbergite). **Component** is a general term which may also be used in systems where phases have fixed compositions.

contact aureole   The zone of metamorphic rocks surrounding an igneous intrusion, which were metamorphosed by the heat of the intrusion.

contact metamorphic rocks   Rocks which underwent metamorphism near the contacts of igneous intrusions, due to the flow of heat out of the intrusion.

contact metamorphism   The metamorphic processes which occur in **contact metamorphic rocks**.

country rocks   The rocks surrounding an igneous intrusion; sometimes applied also to the rocks surrounding a mineral vein or fault.

daughter isotope   An isotope formed by the radioactive decay of an unstable **parent isotope**. Detected in radiometric dating.

decarbonation   Loss of carbon dioxide from carbonate rocks during metamorphism.

dedolomitization   Breakdown of the dolomitic component in dolomite rocks during metamorphism.

degrees of freedom   The number of ways in which a chemical **system** may be changed without altering the number of **phases** present.

dehydration   Loss of water from rocks during metamorphism.

diadochy   The substitution of one ion for another in the atomic structure of a mineral, without a change of structure, e.g. $Mg^{+2} = Fe^{+2}$ in olivine.

diagenesis Lithification of rocks near the Earth's surface, including crystallization of fine-grained rock matrix. Usually regarded as a sedimentary process.

directional fabric See **fabric**.

divariant An association of phases, or **mineral assemblage** with two degrees of freedom, represented by an area on a **phase diagram**.

domain of equilibrium The volume of a metamorphic rock over which the minerals were in thermodynamic equilibrium at a particular time, i.e. the **system** to which the **Phase Rule** applies in metamorphism.

dynamic metamorphic rocks Rocks found near faults, thrusts or meteorite impact sites, in which deformation processes have been dominant during metamorphism.

dynamic metamorphism The processes giving rise to **dynamic metamorphic rocks**. Rock deformation processes dominate over the effects of temperature and pressure increase.

dynamothermal metamorphic rocks See **regional metamorphic rocks**.

eclogite A rock of basic igneous composition whose major minerals are omphacite and garnet. The metamorphic facies defining the conditions of metamorphism of eclogite.

electron-probe microanalysis A non-destructive technique for analysing minerals, where a narrow beam of electrons excites the production of X-rays. Often abbreviated to EPMA.

equilibrium metamorphic fabric The type of texture, or microstructure, in which grain boundary energy has been reduced to a minimum under the conditions of metamorphism. The associated **mineral assemblage** is therefore probably also an equilibrium assemblage.

exsolution The separation of a homogeneous mineral **phase** into different solid phases, usually as temperature falls.

externally buffered The composition of the **metamorphic fluid** is independent of the **mineral assemblage** in the rock, but is controlled from outside (e.g. by the composition of sea water in a submarine hydrothermal circulation system).

external zone A tectonic zone on the outer side of an orogenic belt, adjacent to the stable foreland.

fabric The spatial relationships of the minerals in a rock, such as grain size, grain shape and grain orientation. It is a more general term than **texture** and **microstructure**, because it can be recognized on a larger scale as well as under the microscope.

fibrolite Textural variety of sillimanite, consisting of bundles of prismatic crystals.

field metamorphic gradient The increase in maximum metamorphic temperature with depth in the Earth, or position in a metamorphic terrain. Also known as *metamorphic P–T gradient* or *piezo-thermal array*.

flat   A part of a thrust surface which runs parallel to horizontal layering within the Earth.

flysch   Clastic sedimentary rock sequence deposited **syn-tectonically**.

foliation   A planar metamorphic **fabric**, e.g. **slaty cleavage**, **gneissose banding**.

formation   Lithostratigraphical term for a mappable unit of sedimentary rocks.

garben texture   A rock **fabric** in which prismatic crystals (e.g. of hornblende) are arranged randomly on **foliation** planes.

geotherm   A graph of the increase of temperature with depth within the Earth.

geothermal gradient   The rate of increase of temperature with depth within the Earth, (i.e. the slope of a geotherm).

geothermobarometry   The determination of temperatures and pressures which occurred during metamorphism.

geothermometer   A mineral **system** which may be used to determine (peak) temperatures of metamorphism.

gneiss   A banded metamorphic rock, of medium to high grade, with alternating bands rich and poor in ferromagnesian minerals.

gneissose banding (gneissosity)   A banding caused by metamorphism, with alternating bands rich and poor in ferromagnesian minerals (Fig. 2.19).

Goldschmidt's Mineralogical Phase Rule   The number of minerals in a **mineral assemblage** is equal to, or less than, the number of major oxide components in the rock analysis.

granoblastic texture   Even-grained metamorphic fabric, with planar grain boundaries and 120° triple junctions. Usually a type of **equilibrium fabric**.

granulite   (1) a high grade regional metamorphic rock with abundant anhydrous minerals, especially pyroxenes c.f. **pyroxene gneiss**, (2) quartz- and feldspar-rich, foliated metamorphic rocks.

granulite facies   Low to high P, high T facies of regional metamorphism, in which **granulites** crystallize.

greenschist facies   Medium to high P, low T metamorphic facies of regional metamorphism.

greenstone   A metamorphic rock of basic igneous composition, lacking a cleavage. It typically contains actinolite, epidote and albite. Greenstones have a distinctive pale green colour which distinguishes them from **amphibolite**.

groundmass   The smaller minerals lying between **porphyroblasts**.

group   Lithostratigraphical term for a unit of sedimentary rocks, usually made up of several **formations**.

half life   The time taken for half the unstable **parent isotopes** in a sample to decay.

harzburgite   A variety of peridotite whose major minerals are olivine and orthopyroxene.

hornfels   A massive, often flinty metamorphic rock found nearest to the contact of igneous intrusions.

hydration   Gain of water by rocks during metamorphism.

illite crystallinity   A numerical indication of the ratio of illite to muscovite in white micas of low-grade metamorphic rocks. Used to measure **metamorphic grade** in such rocks.

internally buffered   The composition of the metamorphic fluid is controlled by the mineral assemblage in the rock during metamorphism.

invariant   An association of phases, or **mineral assemblage** with no degrees of freedom, represented by an *invariant point* on a phase diagram.

isochron   A line on an **isochron diagram** whose slope indicates the age of a rock. A line joining the isotopic compositions of different minerals in a rock is a mineral isochron. A line joining different isotopic compositions from several rock samples is called a whole rock isochron.

isochron diagram   A plot of the ratios of **parent** and **daughter isotopes** to a stable isotope (e.g. $^{87}Sr/^{86}Sr$ against $^{87}Rb/^{86}Sr$). The resulting line is an *isochron*.

isograd   A surface joining points of equal **metamorphic grade** in a terrain of metamorphic rocks (which will plot as a line on a geological map). An isograd may be recognized by the appearance of a new index mineral with increasing grade, by the disappearance of a low-grade mineral, or by any other change in the **mineral assemblage**.

isotope   A variety of a chemical element characterized by a particular number of atomic particles in its nucleus e.g. ^{16}O, ^{18}O. Different isotopes of one chemical element have very nearly the same chemical properties, but different atomic masses.

kimberlite   Fragmental volcanic rock containing fragments of rocks from deep in the Earth, sometimes including diamonds.

lepidoblastic   A metamorphic **fabric** in which crystals of platy minerals such as mica lie parallel to one another.

leucosome   The light-coloured part of a **migmatite**, usually of granitoid composition.

lineation   A linear metamorphic **fabric**, such as one in which crystals of prismatic minerals such as hornblende lie parallel to one another.

lithosphere   The outermost layer of the solid Earth, defined by its rigid mechanical properties, which are expressed by relative movements of plates at plate boundaries.

lit-par-lit   A type of migmatite with alternating layers of **leucosome** and **mesosome + melanosome**.

melange   See **tectonic melange**. (Sedimentary melange occurs also.)

melanosome   Part of a migmatite, between **leucosome** and **mesosome** which is enriched in dark ferromagnesian minerals.

mesosome   Part of a migmatite which apparently represents unmelted **protolith**.

metamorphic fabric   See **fabric**.

metamorphic facies   A subdivision of the range of conditions of meta-morphism recognized on the basis of diagnostic **mineral assemblages** or of bounding **metamorphic reactions**.

metamorphic fluid   The fluid phase which is usually present during meta-morphism at grain boundaries and triple junctions and within crystal defects. It may be trapped by growing crystals in fluid inclusions.

metamorphic grade   An approximate measure of the degree of meta-morphism of a rock, often relative, e.g. a rock close to the contact of an intrusion displays *high grade contact metamorphism*, one further away displays *low grade contact metamorphism*.

metamorphic peak   The highest temperature point on the **P–T–t path**.

metamorphism   The sequence of processes which change sedimentary or igneous rocks into metamorphic rocks.

metasomatism   Metamorphism which includes the addition to, or removal from the rock, of chemical **components** other than volatile components such as H_2O and CO_2.

meteoric water   Water which originated at the Earth's surface by falling as rain or snow.

meteor   The streak of light in the upper atmosphere produced by a small **meteoroid** entering the atmosphere and burning up.

meteorite   Part of a **meteoroid** which survives entry to the Earth's atmo-sphere and lands on the surface.

meteoroid   A fragment of solid matter in inter-planetary space, which is attracted by the Earth's gravitational field, and enters the atmosphere.

microstructure   Small-scale structure of a rock, seen by optical or electron microscope. Includes the aspects of grain shape described as **texture**.

migmatite   A mixed rock, often a **gneiss**, with granite mixed with high-grade metamorphic rock.

mineral assemblage   A list of minerals which co-existed in equilibrium during the metamorphism of a rock.

mineral zoning   Variation in composition of individual crystals (often the rim is different from the core).

Moho discontinuity   The sub-horizontal surface separating crust from mantle, within the **lithosphere**.

MORB   'Mid ocean ridge basalt', with distinctive magma composition and igneous petrology. Oceanic basalts may display MORB characteristics without having formed in the *middle* of oceans.

mortar texture   A microfabric in which smaller grains occur along the grain boundaries of larger grains, which are usually strained.

mylonite   Dynamic metamorphic rocks with > 50% of foliated matrix. Usually hard and flinty and found in fault or thrust zones.

nappe   A sheet of rocks transported for a considerable distance over a flat or gently-dipping thrust plane. A large, far-travelled nappe is a type of allochthonous **terrane**.

non-penetrative A non-penetrative process affects some parts of a rock but not others. Thus non-penetrative deformation may give rise to distinct planes of reoriented crystals within a metamorphic rock.

nucleate To begin to grow. Crystals of a new metamorphic mineral may *nucleate* at grain boundaries of pre-existing crystals.

obduction A tectonic process whereby oceanic rocks such as **ophiolites** are uplifted (e.g. by overthrusting) above continental crust.

ocean ridge metamorphism Metamorphic processes occurring beneath oceanic ridges, under conditions where **geothermal gradients** are high, and there is hydrothermal circulation of sea water.

omphacite A green pyroxene, intermediate in composition between augite and jadeite.

ophiolite A sequence of rocks derived from the ocean floor (or part of such a sequence). Ophiolites typically include greenstones, amphibolites and serpentinites.

parent isotope A radioactive isotope which decays, giving rise to a **daughter isotope**.

pelitic rock A metamorphosed sedimentary rock which was originally of mud-rock (shale) composition.

perthite A variety of feldspar in which host crystals of K-feldspar contain exsolved inclusions of plagioclase feldspar.

perturbed geotherm A **geotherm** which displays anomalous features, e.g. inversion of the **geothermal gradient**, because it has been disturbed by tectonic or intrusive events.

phase A physically distinct substance in an experimental system (Fig. 2.6). In an equilibrium **mineral assemblage**, each type of mineral constitutes a phase.

phase change A change in the **phases** in a system, occurring in response to a change in externally-applied conditions, (e.g. a change to denser phases as pressure increases).

phase diagram A diagram representing the states of a chemical system in terms of the **Phase Rule**. If there are two **degrees of freedom** or less, a phase diagram can be drawn on two-dimensional paper. When the degrees of freedom considered are temperature and pressure, a phase diagram can represent the conditions of stability of minerals during metamorphism.

Phase Rule (Gibbs' Phase Rule) A law of physical chemistry relating the number of **phases** in a chemical system to the number of **components** and **degrees of freedom**.

poikiloblastic texture A metamorphic fabric in which large grains of metamorphic minerals enclose smaller grains of other minerals.

polymorph One of several solid **phases** which have the same chemical composition, e.g. andalusite, which is one polymorph of Al_2SiO_5, the others being kyanite and sillimanite.

porphyroblast A larger crystal occurring in a metamorphic rock among a population of smaller crystals, which constitute the **groundmass**. The metamorphic equivalent of a *phenocryst* in igneous rocks.

porphyroblastic texture   The microstructure of a rock containing **porphyroblasts**.

porphyroclast   A fragment of a crystal or of rock, often of lenticular shape, formed by fragmentation during deformation (i.e. **cataclasis**).

post-tectonic   Formed after mountain-building deformation had (mostly) ceased.

pressure solution   The preferential solution of parts of crystals suffering particularly high stresses during deformation. Often gives rise to **spaced cleavage** microstructures.

principle of uniformity   The belief that processes which can be observed in the Earth today formed the same kinds of rocks and geological structures in the past. Summed up in the phrase 'The present is the key to the past'.

prograde   Prograde metamorphism occurs on the part of the **P–T–t path** where temperatures and pressures are increasing, up to the **metamorphic peak**.

progressive metamorphic sequence   A series of metamorphic rocks showing a progressive increase in metamorphic **grade**, without breaks. See **field metamorphic gradient**.

protolith   The rock type that was present before metamorphism.

psammitic rock   A metamorphosed sedimentary rock which was originally of quartz sand or arkose composition.

pseudomorph   A fine-grained aggregate of minerals replacing a pre-existing mineral grain, and preserving its crystal outline.

pseudotachylyte (also spelt 'pseudotachylite')   Glassy rock produced by frictional melting in a fault or thrust zone.

P–T–t path   The sequence of pressures and temperatures which occurred during metamorphism, forming a loop on a temperature–pressure graph.

pyroxene hornfels   Hornfels containing major pyroxene, found in the high temperature parts of **contact aureoles**.

pyroxene hornfels facies   High T facies of ordinary contact metamorphism (maximum temperature approximately the same as that found at the contact of a large intrusion of magma of basic composition).

ramp   The structure formed where a thrust 'steps' upwards through a horizontal layer in the Earth.

regional metamorphic rocks   Metamorphic rocks forming large parts of the Earth's crust or mantle. They formed by the interaction of several metamorphic processes associated with plate tectonics, and do not have a simple genetic relationship to intrusions, faults, or thrusts.

regolith   The fragmental outer layer of the Moon or other planetary bodies. Loosely speaking, 'lunar soil'.

relict mineral   A mineral which survives outside its field of stability, at later times along the **P–T–t path**.

remanent magnetism Permanent magnetism preserved in ferromagnetic minerals in rocks, as they cool past their Curie temperature.

retrograde Retrograde metamorphism occurs on the part of the **P–T–t path** where temperatures are falling from the **metamorphic peak**.

rock cleavage Tendency of a rock to split into sheets. In metamorphic rocks it is usually caused by a **foliation** fabric.

saccharoidal marble Coarse-grained metamorphic rock composed of carbonate minerals (e.g. calcite), with a sugary appearance.

schist Medium to high grade metamorphic rock of pelitic composition, with a well-developed planar **fabric**.

schistosity Tendency of **schist** to split into planar sheets; the planar foliation fabric which causes this tendency.

SEM 'Scanning electron microscope'.

sieve texture A type of **poikiloblastic texture** in which the large crystals are crowded with inclusions of small crystals, displayed by cordierite in **contact aureoles**.

slate A fine-grained pelitic rock displaying slaty cleavage.

slaty cleavage (also spelt 'slatey cleavage') A penetrative foliation fabric in the phyllosilicate minerals of fine-grained pelitic rocks (i.e. slates), which permits the rocks to be split into thin, almost perfectly flat sheets.

smectite A general name for clay minerals of the montmorillonite group. Used in Chapter 5 for clay minerals formed by the low temperature hydration of basaltic glass.

SMOW 'Standard Mean Ocean Water', providing the standard isotopic ratios of D/H and $^{18}O/^{16}O$ used in stable isotope analysis.

solid solution See **diadochy**.

solvus A boundary on a phase diagram (often **univariant**) dividing two distinct *solid* phases.

spaced cleavage Tendency of a rock to split along discrete, closely-spaced fracture planes, a type of **non-penetrative foliation**. Also called *false cleavage*.

spotted slate A contact metamorphic rock, usually of **pelitic** composition, with dark or light spots appearing on the **slaty cleavage** surfaces, which formed during earlier regional metamorphism.

steady-state continental geotherm The **geotherm** reached in the crust and mantle of areas of the lithosphere which have been tectonically stable for a long time (> 100 Ma). Such areas all have continental crust.

strain (1) the change of shape of rocks due to deformation (2) the effects of such a change of shape on individual crystals, such as shadowing of extinction in quartz, and bending of twin lamellae in plagioclase.

suevite A **breccia** of rock fragments in impactite glass.

supergroup Lithostratigraphical term for a large-scale unit of sedimentary rocks, usually containing many **groups** of rocks.

syn-tectonic Formed at the same time as mountain building deformation.

system　An experimentally or theoretically isolated sample of material of simple chemical composition, used for the study of the mutual stability relationships of individual **phases**. Usually described in terms of the formulae of its limiting chemical **components**, e.g. 'the system MgO–SiO_2–H_2O.

tectonic melange　A rock consisting of a mixture of rock types formed by deformation processes. The blocks in the mixture are of varied size and shape, and include *exotic* fragments, not found adjacent to the melange itself.

TEM　'Transmission electron microscopy'.

terrain　A region of metamorphic rocks. e.g. 'The Grampian Highland metamorphic terrain'.

terrane　American spelling of **terrain**, but the term is now used for a far-travelled tectonic unit, with characteristic stratigraphical and structural sequences, more accurately called an *allochthonous terrane* or *suspect terrane*. e.g. 'The Connemara (allochthonous) terrane'.

texture　The relationship between size, shape and orientation of grains in a rock, viewed under the microscope. The meaning overlaps considerably with **microstructure**.

thermal conduction　Transfer of heat by conduction through solids, without displacement of atoms.

thermal conductivity　A measure of the ease with which heat is conducted through rocks.

thermal convection　Transfer of heat by the movement of hot atoms. In metamorphism, the moving atoms are in **metamorphic fluids**.

thermal metamorphic rocks　See **contact metamorphic rocks**.

tie-line　Line joining compositions of minerals which co-existed in equilibrium during metamorphism, on a composition–assemblage diagram.

transform fault　A major fault at a plate boundary, where two plates slide past each other, without extension or compression of the **lithosphere**.

triple junction　The line where three crystals intersect, which appears as a point in thin section. Not to be confused with a **triple point** on a **Phase diagram**, or with *triple junctions* in plate tectonics.

triple point　The temperature and pressure at which three phases are in equilibrium on a phase diagram representing a system with two degrees of freedom, on which it will appear as an invariant point.

univariant　An association of phases, or **mineral assemblage** with one degree of freedom, represented by a line on a phase diagram.

variance　An alternative name for **degrees of freedom**. A system with many degrees of freedom is said to have a high variance.

white mica　A general term for dioctahedral phyllosilicates of the muscovite–illite series.

Wilson Cycle　A summary of the sequence of initial extension of the **lithosphere**, leading to the formation of oceanic crust, followed by subduction and finally continental collision, which has occurred in many orogenic belts. Named for the geophysicist J.T. Wilson.

work hardening   The tendency of many solid substances, including rocks, to increase in strength as they are deformed.

xenolith   A block of country rock surrounded by igneous rock in an intrusion.

References

Anderson, R.N. & J.N. Skilbeck 1981. Oceanic heat flow. In *The Oceanic Lithosphere. (The Sea No. 7)*, C. Emiliani (ed.), 489–524. New York: Wiley.

Ashworth, J.R. (ed.) 1985. *Migmatites*. Glasgow: Blackie.

Bailey, E.B. 1968. *Tectonic essays: mainly Alpine*. (facsimile reprint). Oxford: Oxford University Press.

Barber, A. 1985. A new concept of mountain building. *Geology Today* 1, 116–121.

Barker, A.J. & M.W. Anderson 1989. The Caledonian structural-metamorphic evolution of south Troms, Norway. In *Evolution of metamorphic belts*, J.S. Daly, R.A. Cliff & B.W.D. Yardley (eds), 385–390. Geological Society Special Publication No. 43.

Battey, M.H. 1981. *Mineralogy for students*. London: Longman.

Bell, T.H. & M.J. Rubenach 1980. Crenulation cleavage development – evidence for progressive bulk inhomogeneous shortening from 'millipede' microstructures in the Robertson River Metamorphics. *Tectonophysics* 68, T9–T15.

Berthe, D., P. Choukroune & P. Jegouzo 1979. Orthogneiss, mylonite and non-coaxial deformation of granites: the example of the South Armorican Shear Zone. *Journal of Structural Geology*, 1, 31–42.

Bohlen, S.R. & E.J. Essene 1977. Feldspar and oxide geothermometry of granulites in the Adirondack Highlands. *Contributions to Mineralogy and Petrology* 62, 153–169.

Bohlen, S.R., J.W. Valley & E.J. Essene 1985. Metamorphism in the Adirondacks. I. Petrology, pressure and temperature. *Journal of Petrology* 26, 971–992.

Borradaile, G.J., M.B. Bayly & C.M. Powell 1982. *Atlas of deformational and metamorphic textures*. Berlin: Springer.

Bott, M.H. 1982. *The interior of the Earth: its structure, constitution and evolution*. London: Edward Arnold.

Boullier, A.B. & A. Nicholas 1975. Classification of textures and fabrics of peridotite xenoliths from South Africa kimberlites. *Physics and Chemistry of the Earth* 9, 467–475.

Boyle, A.P. 1985. Metamorphism of basic and pelitic rocks at Sulitjelma, Norway. *Lithos* 19, 113–128.

Boyle, A.P. 1987. A model for stratigraphic and metamorphic inversions at Sulitjelma, central Scandes. *Geological Magazine* 124, 451–466.

Boyle, A.P. 1989. The geochemistry of the Sulitjelma ophiolite and associated volcanics: tectonic implications. In *The Caledonide Geology of Scandinavia*, R.A. Gayer (ed.), 153–164. London: Graham & Trottman.

Brodie, K.H. & E.H. Rutter 1985. On the relationship between deformation and metamorphism, with special reference to the behaviour of basic rocks. In *Metamorphic reactions: kinetics, textures and deformations*, A.B. Thompson & D.C. Rubie (eds), 138–179. New York: Springer.

Brown, G.C. & A.E. Mussett 1981. *The inaccessible Earth*. London: Allen & Unwin.

Burton, K.W., A.P. Boyle, W.L. Kirk & R. Mason 1989. Pressure, temperature and structural evolution of the Sulitjelma fold-nappe, central Scandinavian Caledonides. In *Evolution of Metamorphic Belts*, J.S. Daly, R.A. Cliff & B.W.D. Yardley (eds), 391–411. Geological Society Special Publication, no. 43.

Cheng, Y. 1986. The Proterozoic. In *The Geology of China*, Z. Yang, Y. Cheng & H. Wong (eds). Oxford: Clarendon Press, 31–49.

Chopin, C. 1984. Coesite and pure pyrope in high-grade blueschists of the western Alps: a first record and some consequences. *Contributions to Mineralogy and Petrology* **58**, 255–262.

Cliff, R.A. 1985. Isotopic dating in metamorphic belts. *Journal of the Geological Society of London* **142**, 97–110.

Cox, K.G., J.D. Bell & R.J. Pankhurst 1979. *The interpretation of igneous rocks*. London: Allen & Unwin.

Crawford, M.L. & L.S. Hollister 1986. Metamorphic fluids: the evidence from fluid inclusions. In *Fluid-rock interactions during metamorphism*, J.V. Walther & B.J. Wood (eds), 1–35. Berlin: Springer.

Craig, G.Y. 1983. *Geology of Scotland.* Edinburgh: Scottish Academic Press.

Deer, W.A., R.A. Howie & J. Zussman 1966. *An introduction to the rock-forming minerals*. London: Longman.

Dempster, T.J. 1986. Uplift patterns and orogenic evolution in the Scottish Dalradian. *Journ. Geol. Soc. London* **142**, 111–128.

Dodson, M.H. 1973. Kinetic processes and the thermal history of slowly cooling rocks. *Nature* **259**, 551–553.

Dunoyer de Segonzac, G. 1970. The transformation of clay minerals during diagenesis and low-grade metamorphism: a review. *Sedimentology* **15**, 281–346.

Eastwood, T., S.E. Hollingworth, W.C.C. Rose & F.M. Trotter 1968. *Geology of the country around Cockermouth and Caldbeck. (Explanation of One-inch Geological Sheet 23, New Series)*. London: Institute of Geological Sciences.

Elthon, D. 1981. Metamorphism in oceanic spreading centers. In *The Oceanic Lithosphere. (The Sea No. 7)*, C. Emiliani (ed.), 285–304. New York: Wiley.

England, P.C. & S.W. Richardson 1977. The influence of erosion upon the mineral facies of rocks from different metamorphic environments. *Journ. Geol. Soc. London* **134**, 201–213.

England, P.C. & A.B. Thompson 1984. Pressure-temperature-time paths of regional metamorphism. Part I. Heat transfer during evolution of regions of thickened continental crust. *Journal of Petrology* **25**, 849–928.

Eskola, P. 1920. The mineral facies of rocks. *Norsk Geol. Tidsskr.* **6**, 143–194.

Etheridge, M.A., V.J. Wall & R.H. Vernon 1983. The role of the fluid phase during regional metamorphism and deformation. *Journal of Metamorphic Geology* **1**, 205–226.

Evans, A.M., T.D. Ford & J.R.L. Allen 1968. Precambrian Rocks. In *Geology of the East Midlands*, P.C. Sylvester-Bradley & T.D. Ford (eds), 1–19. Leicester: Leicester University Press.

Evans, D.J. & S.H. White 1984. Microstructural and fabric studies from the rocks of the Moine Nappe. *Journal of Structural Geology* **6**, 369–389.

Faure, G. 1986. *Principles of isotope geology*. New York: Wiley.

Ferguson, J., H. Martin, L.O. Nicolaysen & R.V. Danchin 1975. Gross Brukkaros: a kimberlite-carbonate volcano. *Physics and Chemistry of the Earth* **9**, 219–234.

Firman, R.J. 1978. Intrusions. In *Geology of the Lake District*. F. Moseley (ed.), 146–163. Yorkshire Geological Society Occasional Publication, no. 3.

Frey, M. 1987. *Low temperature metamorphism*. Glasgow: Blackie.

Frey, M. & J.C. Hunziker 1973. Progressive niedriggrade Metamorphose glauconiführendender Horizonte in den helvetischen Alpen der Ostschweiz. *Contr. Mineral. Petrol.* **39**, 185–218.

Fry, N. 1984. *The field description of metamorphic rocks*. Milton Keynes: Open University Press.

Fyfe, W.S., F.J. Turner & J. Verhoogan 1958. *Metamorphic reactions and metamorphic facies*. Geological Society of America, Memoir 73.

Gass, I.G. & J.D. Smewing 1981. Ophiolites: obducted oceanic lithosphere. In *The Oceanic Lithosphere. (The Sea, No. 7)*, C. Emiliani (ed.), 339–362. New York: Wiley.

Gill, R. 1989. *Chemical fundamentals of geology*. London: Unwin Hyman.

Goldschmidt, V.M. 1911. *Die Kontaktmetamorphose im Kristianiagebiet*. Vidensk. Skrifter I. Mat.-Naturv. K (1911), No. 11.

Gribble, C.D. 1988. *Rutley's elements of mineralogy*. London: Unwin Hyman.

Gribble, C.D. & A.J. Hall 1985. *A practical introduction to optical mineralogy*. London: Allen & Unwin.

Hall, R. 1976. Ophiolite emplacement and the evolution of the Taurus suture zone, southeastern Turkey. *Bull. Geol. Soc. Amer.* **87**, 1078–1088.

Hall, R. 1980. Unmixing a melange: the petrology and history of a disrupted and metamorphosed ophiolite, SE Turkey. *Journ. Geol. Soc. London* **137**, 195–206.

Hansen, E.C., R.C. Newton & A.S. Janardhan 1984. Fluid inclusions in rocks from the amphibolite-facies to charnockite progression in southern Karnataka, India: direct evidence concerning the fluids of granulite metamorphism. *Journal of Metamorphic Geology* **2**, 249–264.

Harker, A. 1932. *Metamorphism*. London: Methuen.

Harte, B. 1983. Mantle peridotites and processes. In *Continental basalts and mantle xenoliths*, C.J. Hawkesworth & M.J. Norry (eds), 46–91. Nantwich, Cheshire: Shiva.

Harte, B. & J.J. Gurney 1975. Evolution of clinopyroxene and garnet in an eclogite nodule from Roberts Victor kimberlite pipe, South Africa. *Physics and Chemistry of the Earth* **9**, 367–387.

Hedberg, H.D. (ed.) 1976. *International stratigraphic guide*. New York: Wiley.

Hobbs, B.E., W.D. Means & P.F. Williams 1976. *An outline of structural geology*. New York: Wiley.

Hutton, D.H.W. & J.F. Dewey 1986. Palaeozoic terrane accretion in the western Irish Caledonides. *Tectonics* **5**, 1115–1124.

Jaeger, J.C. 1957. The temperature in the neighborhood of a cooling intrusive sheet. *American Journal of Science* **255**, 306–318.

Jeanloz, R. & A.B. Thompson 1983. Phase transitions and mantle discontinuities. *Reviews of Geophysics & Space Physics* **21**, 51–74.

Jones, D.L., N.J. Silberling, W. Gilbert & P.J. Coney 1982. Character, distribution, and tectonic significance of accretionary terranes in the central Alaska range. *Journal of Geophysical Research* **87**, 3709–3717.

Kirk, W.L. & R. Mason 1984. Facing structures in the Furulund Group, Sulitjelma, Norway. *Proceedings of the Geological Association* **95**, 43–50.

Knipe, R.J. & R.P. Wintsch 1985. Heterogeneous deformation, foliation development and metamorphic processes in a polyphase mylonite. In *Metamorphic reactions: kinetics, textures and deformations*, A.B. Thompson & D.C. Rubie (eds), 180–210. New York: Springer.

Lappin, M.A. & J.B. Dawson 1975. Two Roberts Victor eclogites and their re-equilibration. *Physics and Chemistry of the Earth* **9**, 351–366.

Leliwa-Kopystynski, J. & R. Teisseyre 1984. *Constitution of the Earth's interior*. Amsterdam: Elsevier.

Ma, X.Y. & Z.W. Wu 1981. Early tectonic evolution of China. *Precambrian Research* **14**, 185–202.

Mason, R. 1971. The chemistry and structure of the Sulitjelma gabbro. *Norges Geologiske Undersokelse*, no. 269, 108–141.

Mason, R. 1985. Ophiolites. *Geology Today* **1**, 136–140.

Mason, R. 1988. Did the Iapetus Ocean really exist? *Geology* **16**, 823–826.

Mason, R. 1989. Teaching Geology in China. *Geology Today* **5**, 102–104.

McKenzie, D.P. 1978. Some remarks on the development of sedimentary basins. *Earth and Planetary Science Letters* **40**, 25–32.

McKenzie, W.S., C.H. Donaldson & C. Guildford 1982. *Atlas of Igneous Rocks and their Textures*. London: Longman.

McKenzie, W.S. & C. Guildford 1980. *Atlas of rock-forming minerals in thin section*. London: Longman.

Means, W.D. 1976. *Stress and strain*. New York: Springer.

Miyashiro, A. 1961. Evolution of metamorphic belts. *Journal of Petrology* **2**, 277–311.

Miyashiro, A. 1973. *Metamorphism and Metamorphic Belts*. London: Allen & Unwin.

Morrison, J. & J.W. Valley 1988. Post-granulite facies fluid infiltration in the Adirondack Mountains. *Geology* **16**, 513–516.

Moseley, F. 1978. *Geology of the Lake District*. Yorkshire Geological Society Occasional Publication, no. 3.

Nicholas, A. & J.P. Poirier 1976. *Crystalline plasticity and solid state flow in metamorphic rocks*. New York: Wiley.

Nicholson, R. 1966. On kink-zone development and metamorphic differentiation in the low-grade schists of Sulitjelma. *Norges Geologiske Undersokelse* **247**, 133–145.

Nixon, P.H. (ed.) 1987. *Mantle xenoliths*. New York: Wiley.

Oxburgh, E.R. & D.L. Turcotte 1974. Thermal gradients and regional metamorphism in overthrust terrains with special reference to the eastern Alps. *Schweiz. Mineral. und Petrogr. Mitt.* **54**, 641–662.

Peter, W.C. 1987. *Exploration and mining geology* New York. Wiley.

Piper, J.D.A. 1974. The Sulitjelma gabbro, Norway: a palaeomagnetic result. *Earth and Planetary Science Letters* **21**, 383–388.

Piper, J.D.A. 1987. *Palaeomagnetism and the continental crust*. Milton Keynes: Open University Press.

Potts, P.J. 1987. *A handbook of silicate rock analysis*. Glasgow: Blackie.

Powell, C.M. & R.H. Vernon 1979. Growth and rotation history of garnet porphyroblasts with inclusion spirals in the Karakorum schist. *Tectonophysics* **54**, 25–43.

Rastall, R.H. 1910. The Skiddaw granite and its metamorphism. *Quart. Journ. Geol. Soc. London* **66**, 116–141.

Read, H.H. 1957. *The Granite Controversy*. London: Thomas Murby.

Reid, A.M., C.H. Donaldson, J.B. Dawson, R.W. Brown & W.I. Ridley 1975. The Igwisi Hills extrusive 'kimberlites'. *Physics and Chemistry of the Earth* **9**, 199–218.

Ronov, A.B. & A.A. Yaroshevshkiy 1976. A new model for the chemical structure of the Earth's crust. *Geochem. Inst.* **13**, 89–121.

Schmid, S.M. 1975. The Glarus overthrust: field evidence and mechanical model. *Eclogae geol. Helveticae* **68**, 247–280.

Selley, R.C. 1985. *Elements of petroleum geology*. New York: Freeman.

Şengör, A.M.C. & Y.Yimaz 1981. Tethyan evolution of Turkey: a plate tectonic approach. In *Tectonophysics* **75**, 181–241.

Sibson, R.H. 1977. Fault rocks and fault mechanisms. *Journ. Geol. Soc. Lond.* **133**, 191–213.

Sobolev, V.S. (ed.) 1972. *The facies of metamorphism*, (translated by D.A. Brown), Canberra: Australian National University. (Original Russian edition 1970. Moscow: Nedra.)

Soper, N.J. 1987. The Ordovician batholith of the English Lake District. *Geol. Mag.* **124**, 481–482.

Soper, N.J. & D.E. Roberts 1971. Age of cleavage in the Skiddaw Slates in relation to the Skiddaw aureole. *Geol. Mag.* **108**, 293–302.

224 REFERENCES

Spear, F.S. & J. Selverstone 1983. Quantitative P–T paths from zoned minerals: theory and tectonic applications. *Contributions to Mineralogy and Petrology* **83**, 348–357.

Spear, F.S., J. Selverstone, D. Hickmott, P. Crowley & K.V. Hodges 1984. P, T paths from garnet zoning: a new technique for deciphering tectonic processes. *Geology* **12**, 87–90.

Spry, A. 1969. *Metamorphic textures*. Oxford: Pergamon Press.

Stanton, R.L. 1972. *Ore petrology*. New York: Wiley.

Stephens, M.B. 1988. The Scandinavian Caledonides: a complexity of collisions. *Geology Today* **4**, 20–26.

Stoffler, D. 1966. Zones of impact metamorphism in the crystalline rocks of the Nordlinger Ries Crater. *Contributions to Mineralogy and Petrology* **12**, 15–24.

Sutton, J. & J. Watson 1951. The pre-Torridonian metamorphic history of the Loch Torridon and Scourie areas in the North-west Highlands. *Quart. Journ. Geol. Soc. Lond.* **106**, 241–307.

Taylor, H.P. 1974. The application of oxygen and hydrogen isotope studies to problems of hydrothermal alteration and ore deposits. *Economic Geology* **69**, 843–883.

Taylor, H.P. & R.W. Forester 1981. Low O^{18} igneous rocks from the intrusive complexes of Skye, Mull and Ardnamurchan, western Scotland. *Journal of Petrology* **12**, 465–497.

Thompson, A.B. 1981. The pressure-temperature (P, T) plane viewed by geophysicists and petrologists. *Terra Cognita* **1**, 11–20.

Thompson, A.B. 1983. Fluid-absent metamorphism. *Journ. Geol. Soc. London* **143**, 533–549.

Thompson, J.B. 1957. The graphical analysis of mineral assemblages in pelitic schists. *Am. Mineral.* **42**, 842–858.

Thompson, P.H. 1977. Metamorphic P–T distributions and the geothermal gradients calculated from geophysical data. *Geology* **5**, 520–522.

Tilley, C.E. 1924. The facies classification of metamorphic rocks. *Geol. Mag.* **61**, 167–171.

Turner, F.J. 1981. *Metamorphic petrology: mineralogical, field and tectonic aspects*. New York: McGraw Hill.

Vogt, T. 1927. *Sulitelmafeltets geologi og petrografi*. Norges Geologiske Undersokelse no. 121. (English summary.)

Walther, J.V. & B.J. Wood 1984. Rate and reaction in prograde metamorphism. *Contributions to Mineralogy and Petrology* **88**, 246–259.

Warner, J.L. 1972. Metamorphism of Apollo 14 breccias. *Proc. Third Lunar Sci. Conf.* (Suppl. 4 to *Geochim. et Cosmochim. Acta*), **1**, 481–504.

Watkins, K.P. 1983. Petrogenesis of Dalradian albite porphyroblast schists. *Journ. Geol. Soc. London* **136**, 663–671.

Wilson, J.T. 1966. Did the Atlantic close and then re-open? *Nature* **211**, 676–681.

Wilson, M.R. 1972. Strain determination using rotational porphyroblasts, Sulitjelma, Norway. *Journal of Geology* **80**, 421–431.

Windley, B.F. 1984. *The evolving continents*. New York: Wiley.

Winkler, H.G.F. 1986. *Petrogenesis of metamorphic rocks*. New York: Springer.

Yang, Z.Y., Y.Q. Cheng & H.Z. Wang 1986. *The geology of China*. Oxford: Clarendon Press.

Index

Q'shot
e10990402
£11.50